Injection stretch-blow moulding companion

This book belongs to

Bob Blakeborough

© Bob Blakeborough 2014
Third edition

The moral right of the author has been asserted.

All rights reserved. Without limiting the rights under copyright reserved above, no part of this publication may be reproduced, stored in or introduced into a retrieval system, or transmitted, in any form or by any means (electronic, mechanical, photocopying, recording, or otherwise), without the prior written permission of the author.

ISBN 1-794-81527-8

Contents

Safety 7

1. Introducing container production
   How and why injection stretch-blow moulding was developed 9

2. Injection stretch-blow moulding
   The process, and the one-stage machine 13

3. Materials
   Thermoplastics, PET, IV, AA, preform design. Other materials 29

4. Drying
   Drying knowhow 41

5. Hydraulics
   Components, symbols and circuits 47

6. Pneumatics
   Components, symbols and circuits 63

7. Electrics
   Components and symbols 69

8. Plant engineering
   Principles and factory layout considerations 79

9. Quality Assurance
   Typical measurement techniques, test methods and statistical analysis 89

10. Process optimisation
    A strategy for establishing the optimal process 101

11. Troubleshooting
    Strategies, a logical approach, correcting process faults 115

12. Glossary 121

Useful information 127
   Conversion Factors, Density of Water and Neck finish terminology

Index 131

Safety

Legal obligations

All workers have a right to work in places where risks to their health and safety are properly controlled. Although the overall responsibility for this may be down to the employer, under the provisions of the UK's Health and Safety at Work Act both workers and employers have a legal responsibility to look after health and safety at work together. Every employee, whether they are permanent staff, agency, or a contractor is required:

- to take reasonable care of their own health and safety
- to take reasonable care of the health and safety of anyone who may be affected by their acts (or omissions)
- to co-operate with their employer or any other person to enable legal obligations to be met
- not to misuse or interfere with anything provided in the interests of health and safety

Critical incidents

Research has shown that for every event resulting in major injury there may have been as many as ten which resulted in minor injury, thirty causing property damage, and hundreds in which the outcome was neither injury nor damage. The point is that any critical incident might result in personal injury; the seriousness of the outcome is purely a matter of chance. Therefore should you come across a potentially dangerous situation and are not in a position to deal with it immediately yourself, report your concern without delay to the appropriate authority.

Pressurised systems

Electrical systems are widely recognized as being potentially lethal (the lack of any visible sign that a conductor is electrified being a particular hazard). Hydraulic and pneumatic systems are no less dangerous; but tend to be approached in a far more carefree manner. High pressure air or oil released suddenly can readily result in injury. Unexpected movement of components such as cylinders can trap and crush limbs. Spilt hydraulic oil is likely to lead to falls and injury. It follows that hydraulic and pneumatic systems should be treated with respect, and maintained or repaired under well-defined procedures and safe working practices as rigorous as those applied to electrical equipment.

1. Before starting work, think of the implications of what you're about to do, and make sure anyone who could be affected knows of your intentions.
2. Ensure anything that can move with changes in pressure/ as a result of your actions is mechanically secured or guarded.
3. When working with hydraulic systems make prior arrangements to catch oil spillage. Have containers and cloths ready, and as far as is possible, keep spillage off the floor. Clean up any spilt oil as soon as possible.
4. When the job is finished check for leaks and correct operation. Leave the area tidy and clean. Ensure other workers know that machinery is about to move again.

Heat

The high temperatures and pressures inside the injection barrel can result in a violent explosion of hot plastic from the nozzle if proper procedures are not followed. Hot plastic adheres to skin causing severe burns. Therefore during purging wear protective clothing, ensure the guards are in place, and always use low operating pressures. When the machine is not operating, retract the injection unit from the hot runner system so that any pressure build up can be relieved through the sprue bush (inlet or hard radius). Always reduce temperature settings if the machine is not to be restarted immediately.

Safety Guards and interlocks

Moving machine parts have the potential to cause severe injury and so guards are provided to control access. Remember that safety guards are provided for your protection. Respect them, and do not attempt to bypass them or override safety interlocks.

Introducing plastic container production

Blow moulding involves placing a mass of molten material inside a closed mould and inflating it so that it assumes the shape of the cavity. Once the material has cooled, the mould is opened and the resultant *container* removed. Glass bottles have for centuries been made by blow moulding. In this case a gather of molten glass collected on the end of a blowpipe is blown inside a bottle-shaped mould. Blow moulding is only feasible with materials such as glass that have an intermediate *plastic* phase between the solid and liquid states (unlike the abrupt transition between liquid water and solid ice). Thermoplastic materials have a similar intermediate plastic phase and thus can also be blow moulded. There are three different plastic container blow moulding processes:

- Extrusion Blow Moulding (EBM)
- Injection Blow Moulding (IBM)
- Injection Stretch-Blow Moulding (ISBM)

Extrusion blow moulding can produce hollow parts ranging in size from large septic tanks to small eye-drop squeeze bottles. Initially a molten open-ended tube of resin is extruded through an annular die and captured between two blow mould halves as they close. Mating surfaces on the mould pinch together the open areas of the tube and, since the material is hot enough to *weld*, seal the lower end to produce a *parison*. Compressed air is then introduced through a needle to inflate the parison. Where the distended parison touches the blow mould surface, it cools and hardens to form a hollow container of the same shape as the cavity.

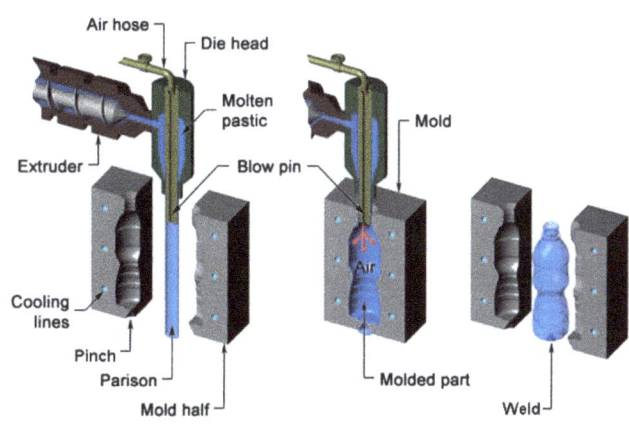

Extrusion Blow Moulding

After cooling, the mould is opened, the part is ejected, and the excess material is trimmed from the top and bottom pinch-off areas. In this manner, blow moulded parts, including bottles for household and food products (e.g. milk) are economically produced. While HDPE is the largest volume thermoplastic used in extrusion blow moulding, containers are also routinely made from PP, LLDPE, LDPE, and PVC.

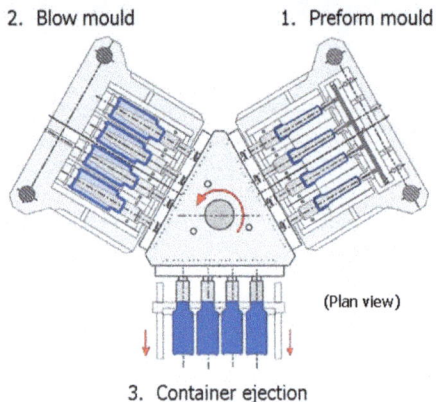

Injection Blow Moulding

In the **Injection Blow Moulding** process, a preform is injection moulded around a core pin. The hot preform, still on the core pin, is then indexed to a blow moulding station where it is blown and cooled, before at a third station the container is stripped from the core pin. Injection blow moulded containers do not need to be trimmed, greatly reducing the amount of scrap. In addition, the process provides precision detail in the neck and thread finish, and it affords an easy and reproducible method of optimizing wall thickness. Unlike stretch blow moulding, this process does not afford much in the way of orientation or provide any improvement in properties. Further, its use is normally limited to the production of relatively small thick-wall containers. Again HDPE, LDPE, PP, and PVC are the most commonly moulded materials, but other possibilities include PS, SAN, PC, and PET.

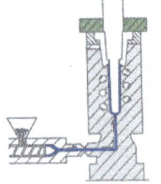  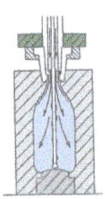 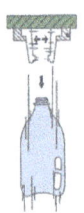

Injection   (Conditioning)   Blowing   Ejection

The **Injection Stretch-Blow Moulding** process involves the production of thin-wall containers having biaxial molecular orientation. This biaxial orientation provides enhanced physical properties, clarity, and gas barrier properties; all vital characteristics in bottles intended to hold carbonated soft drinks.

There are two distinct stretch blow moulding techniques, one-stage and two-stage. In the one-stage process the preforms are injection moulded, conditioned to the proper temperature, and blown into containers in one continuous process. This technique is most effective for specialty or custom applications, e.g. wide mouth jars, where production quantities are typically lower.

In the two-stage process, preforms are injection moulded and may be stored before they are blown into containers on a separate reheat blow (RHB) machine. Because of the relatively high cost of the injection moulding and RHB equipment, this technique is generally used for high volume applications such as carbonated soft drink bottles.

How and why was the injection stretch-blow moulding process developed?

Until the 1970s, almost all carbonated soft drinks (CSD) were packaged in metal cans or small returnable glass bottles. Attempts to introduce family-size glass bottles had been short-lived due to their heavy weight and the dire consequences of breakage. What was needed was a plastic that could be used to make cost-effective, lightweight, family-size bottles for carbonated soft drinks.

Weight of 64 oz glass bottle:

container: 1.8 kg

beverage: 1.9 kg

total: 3.7 kg [8 lbs]

The bottle-making plastic materials available at the time such as HDPE were unsuitable for two reasons: their low tensile strength, and a gas barrier insufficient to retain the carbon dioxide fizz for a viable shelf life. High tensile strength was needed because the carbon dioxide gas that is dissolved under pressure in a carbonated soft drink pushes outwards against the walls of the container, and the more carbon dioxide that is dissolved (or the higher the ambient temperature) the greater the pressure that it exerts. Glass bottles being rigid, withstand these expansion forces without visible deformation, but an HDPE bottle will *creep* and swell up like a balloon. Any plastic used to make a CSD bottle must have sufficient strength to resist these expansion forces.

The reason that PET succeeded where other plastic materials did not is that PET has a very special characteristic: providing it is heated to a temperature where its chain-like molecules are sufficiently mobile to *unfold* instead of breaking as the material is stretched, it can be *oriented* (stretching applied from two directions at right angles is termed *biaxial* orientation). Oriented PET consists of closely packed molecular chains aligned in the directions of stretch, and is actually stronger than the non-oriented *amorphous* material because the molecules work together to support loads that would break individual chains. Not only is the tensile strength of oriented PET several times that of the amorphous material, the impact strength, gas barrier and chemical resistance are also improved, meaning that bottles made from biaxially oriented PET can be lighter without sacrificing performance.

Traditional extrusion blow moulding techniques were found to be unsuitable for producing biaxially oriented PET containers, partly due to the low cohesive strength of PET when in the melt state, but more crucially, biaxial orientation only occurs in a narrow temperature band that is well below the temperature at which an extruded PET tube will "pinch weld" to form a parison. Since the injection blow moulding process didn't allow sufficient stretching to induce orientation, in the early 1970's the two step injection stretch-blow moulding process was developed. Here a test tube-like preform is injection moulded in a first step, and then in a second step the preform is re-heated until it is sufficiently pliable to be inflated inside a blow mould to form a bottle.

This is known as the two-stage process to differentiate it from the one-stage process swiftly developed by Nissei Plastic Industrial who found that if the preform was ejected from the mould while still hot as in the IBM process, it could be blown without re-heating. In December 1975 Nissei Plastic Industrial unveiled the first commercial machine for making PET bottles: the one-stage ASB-150 injection stretch-blow moulding machine.

The successful introduction of PET bottles led to a rapid rise in demand for machines to produce them, and so two specialist companies were now established to design and to manufacture stretch-blow moulding machines. First Aoki Technical Laboratory Inc. in March 1976, and then Nissei ASB Machine Co., Ltd. in November 1978.

All injection stretch-blow moulding machines deriving from this original "Aoki Stretch Blow" design are referred to in this book as classic one-stage machines. The main distinguishing features of the classic one-stage machine are the *vertical press* to clamp the injection mould, and the *rotary table* beneath which are carried full sets of Lip Cavities (neck splits) for each station.

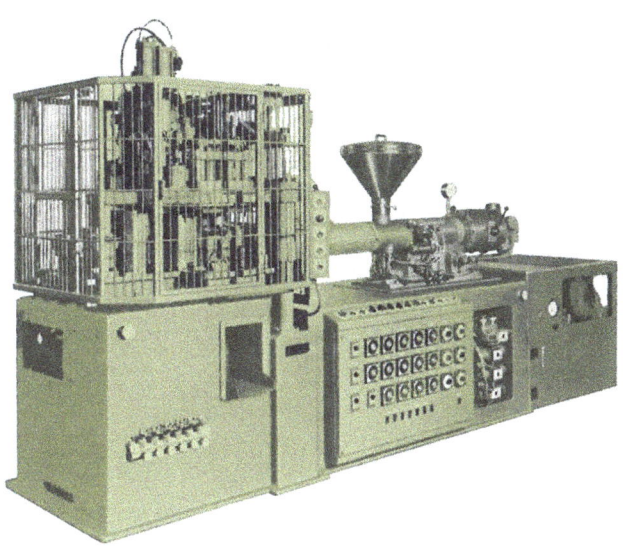

The injection stretch blow-moulding process has several distinct advantages:

- it is energy efficient—using, rather than wasting the heat retained in the preform from the injection moulding process
- biaxial orientation enables optimum material properties to be achieved—meaning bottles can be lighter without sacrificing performance
- since preforms are not subjected to handling, they remain free of surface scratches or scrapes
- the injection moulded preform provides a high-precision neck finish that requires no additional finishing operations; meaning the blown container is ready for decorating/ filling when it exits the machine
- the process does not generate scrap with every container produced

The classic one-stage machine design does have some disadvantages:

- a one-stage machine has the same number of blow moulds as it has injection cavities, yet injection moulding the preform is by far the most time consuming part of the process, as it takes three or sometimes four times as long to cool down as it does to blow; which means that for a part of every cycle, the blow moulds remain idle.
- the preforms/ bottles are supported at each *station* by the *Lip Cavities* (neck splits), meaning a full set is required *at each one*, rather than the single set of lip cavities found in a typical injection tool.

2

Injection stretch-blow moulding

The injection stretch-blow moulding process is akin to a recipe for making a bottle: *Start by injection moulding a preform but extract it from the mould while it still retains sufficient heat to behave like rubber, then — having first ensured the preform is at the correct temperature[1]— place it inside a temperature-controlled blow mould and inflate it until the swelling preform touches the mould surface. Maintain the blow air pressure while the preform cools to form a rigid plastic container with a precise injection moulded neck. Finally exhaust the blow air and remove the finished bottle from the mould.*

The different equipment needed to complete each step of this process is combined in the classic one-stage machine. At its hub is an electrically driven or hydraulically actuated *rotary table* bearing the *lip cavities* (a complete set for each station) which both form the neck and carry the preforms from one station to the next. There will be three or four stations depending upon the machine manufacturer: all machines have *injection*, *blowing*, and *take-out* stations, but some machines also have a *conditioning* station where the temperature of the preforms can be adjusted prior to blowing. All three (or four[1]) production steps take place simultaneously as each set of preforms progresses from one station to the next (represented by the *columns* of the Table).

Injection Station	(Condition Station)[1]	Blow Station	Eject Station
Upper mould close		Blow Core down	
Lower mould close	Conditioning Core down	Blow mould close	Ejector down
Injection (filling)	Conditioning Pot up	Stretch rod down & Blow	
Injection (packing)	Optional gate cut	Exhaust	Ejector up
Curing & Screw charging	Conditioning Pot down	Stretch rod & Blow Core Up	
Upper mould open	Conditioning core up	Blow mould open	
Lower mould open			

The injection station is in two parts: the injection unit, and the clamping unit. The injection unit both melts the plastic pellets and injects the molten material into the cavities, while the vertical hydraulically-actuated press opens and closes the injection mould at high speed. It also applies the *clamp tonnage* to hold the mould shut against the opening force generated by the injection pressure transmitted through the molten polymer. Once the the preform has cooled, the opening of the mould draws the preforms up and out of the cavities.

---

[1] Four-station machine

The conditioning station on a four-station machine provides the facility for individual preform temperature adjustment. It consists of heated pots that are able to rise up from below to surround the preforms, plus heated cores that can be inserted into the preforms from above.

A three-part temperature-controlled aluminium blow mould is mounted in the blow station, and descending from above (piercing the rotary table) a tightly fitting *Blow Core* seals inside the preform neck. Both the stretch rod and the blow air enter the preform via the hole through the centre of each blow core.

At the take-out station the lip cavities are forced apart by pneumatically-actuated wedges, at the same time as a simple centring device holds the container in the mid position to ensure that it will drop out cleanly.

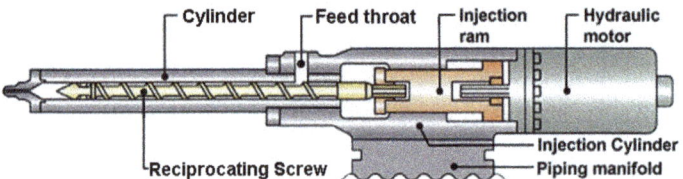

Injection unit

The injection unit consists of a *reciprocating screw* which is rotated inside a heated *cylinder* powered either by an in-line or a radial-piston hydraulic (or sometimes an electric) motor. Pellets falling into the feed throat from the material hopper mounted directly above are conveyed towards the front of the cylinder by the rotation of the screw (as in an auger-conveyor). As the pellets move forward along the screw they are compressed and the combination of conducted, frictional, and compression-induced heating causes them to melt.

As the molten material accumulates at the front of the cylinder, it pushes the screw backwards until the point previously set by the operator is reached which means the desired shot weight has collected in front of the screw, and rotation (and thus charging) may be automatically halted. Next, once the mould has closed and the cylinder nozzle is firmly pressed against the *Sprue Bush* (the inlet to the hot runner system), the Injection Ram thrusts the screw forward like a piston inside a cylinder so that it injects the molten material through the hot runner system and via the *gate* orifice, into the cavity.

To prevent the melt flowing back along the screw during injection, a sliding non-return valve or *Check Ring* is fitted on the front. The screw tip or *torpedo* is of the same shape as the inside of the cylinder head so that the melt remaining in front of the screw when it is fully forward is kept to a minimum.

The material is injected using high pressure for speed and consistency, but once the melt has completely filled the cavities, unrelenting application of high pressure would likely overcome the *clamp tonnage* and force open the mould. To forestall this the high pressure is reduced to a lower *packing* pressure just sufficient to push enough additional material into the cavity to offset the contraction of the cooling plastic (a significant problem in the case of crystalline polymers such as polypropylene which markedly reduce in volume when they crystallize). An additional benefit of applying a reduced packing pressure is that it enables lower clamping forces to be used, thereby allowing the machine to be smaller, lighter, and arguably less expensive than might otherwise have been the case.

The packing pressure has to be maintained until the gates close (precise timing is important as the applied pressure clearly won't be effective after the gates have closed, while stopping it before they close results in variable mouldings). Note that if the screw were to "bottom out" during packing, no more material would be pushed into the cavity; meaning it is crucial to ensure that even at the end of the packing phase, a *cushion* of melt remains in front of the *torpedo*.

While the preform is cooling inside the cavity, the screw rotates again to feed molten material to the front of the screw in readiness for the next shot. Ideally this *charging* will be completed just before the next injection stroke is due to start, so that there is no unnecessary waiting, and the material *residence time* in the cylinder is kept to a minimum. (Polymer that has been kept hot for too long degrades, reducing its physical properties).

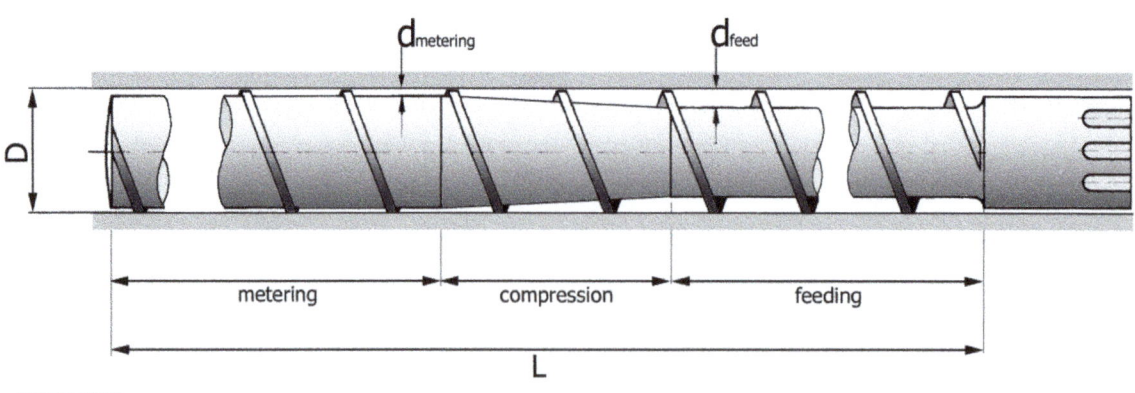

NOT TO SCALE

Screw design:

Injection screws are made from alloy steel that has been hard chrome plated for abrasion and corrosion resistance. They are rated by their L/D (Length/Diameter) Ratio and by their Compression Ratio. For example a "54 mm" diameter screw with an L/D ratio of 19:1 has a diameter of just under 54 mm and is 19 x 54 = 1026 mm long. (The nominal size of the screw relates to the internal diameter of the cylinder; the actual screw diameter being fractionally smaller to give a clearance fit inside the cylinder).

Many manufacturers offer a choice of screw (and mating cylinder) diameters with each injection unit. The screw diameter directly affects not only the L/D ratio but also the injection volume (and hence the shot weight). A larger diameter screw will provide:

(a) a larger injection capacity
(b) but a lower injection pressure
(c) and greater melt temperature variation (due to the increase in shear heating)
(d) as well as reduced energy efficiency.

L/D ratio	Application:
18:1 (or lower)	used for low precision applications where shot weight is the more dominant selection criterion;
20:1	general purpose screw and typically used for ISBM;
22:1 (and above)	provides better mixing and more uniform heating for high precision applications and when using engineering thermoplastics.

In order to better perform its functions of feeding, melting, mixing, and injecting the melt, injection screws have three distinct zones distinguished by their respective flight depths. The deepest at the rear is called the *feed* section, the middle transitional part is called the *compression* section, while the shallow front zone is called the *metering* section. The flight depth in the transitional part gradually tapers from the greater depth in the feed section, to the shallower metering section depth.

Compression Ratio = $d_{feed} / d_{metering}$

Generally one-stage machine injection screws have a *compression ratio* that is appropriate for PET of about 2.5 : 1 (i.e. a little less than the general purpose screws that are typically installed in injection moulding machines). If the compression ratio was excessive, too much heat would be induced in the material, causing thermal degradation. Conversely if the compression ratio was too low, the melt might not be hot enough to eliminate all of the crystallinity in the PET pellets (which would be a serious problem, since any crystalline remnants will act as a nucleating agent, leading to more crystallization and in the worst case, hazy preforms).

The rotation of the screw not only conveys the pellets dropping into the cylinder feed throat towards the front of the cylinder, but as the *flight depth* gets smaller the pellets are compressed and therefore heat up and melt. It is not always appreciated that much of the energy supplied to melt the plastic is mechanical energy induced by the compression and shearing of the plastic. Heat conducted from the heater bands is necessary for the initial melting of the material but is less important after production start-up. Melt temperatures are generally higher than the set temperatures because of this shear heating, and so fast cycling machines with a high throughput of material are fitted with cylinder cooling fans which may be switched on to prevent overheating.

The flight depth in the metering zone needs to be shallow enough to prevent pellets from passing through, but not so shallow that it causes excessive shear heating. A possible solution is to use a *Barrier*

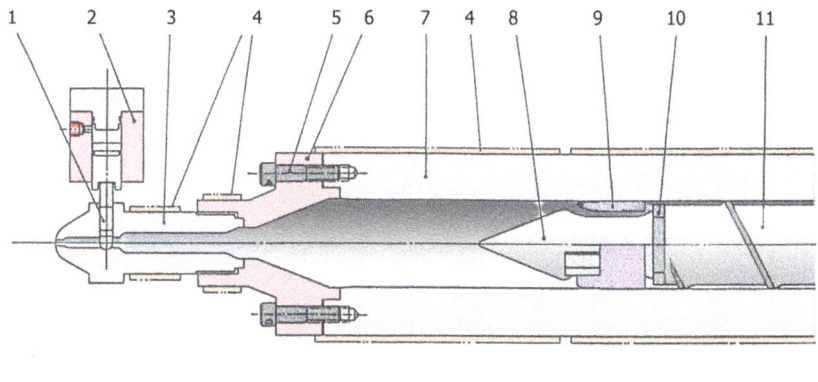

1. Hydraulic Shut-Off Nozzle Pin
2. Hydraulic Shut-Off Nozzle Cylinder
3. Barrel Nozzle
4. Heater Bands
5. Barrel Head fixing bolt
6. Barrel Head
7. Barrel
8. Torpedo
9. Check Ring
10. Seat Ring
11. Injection Screw

Shut-Off Nozzle & Check Ring

Screw designed to keep any pellets that are still solid separate from the already melted material. Barrier screws have a conventional feed section for solids conveying, but in the transitional compression zone have a second supplementary flight — lower than the main flight so that melt, but not pellets can flow over it — separating the melt from the remaining solid material. The narrowing solids channel forces unmelted pellets against the cylinder wall causing efficient frictional melting, while the extra deep melt channel generates less shear, thereby minimizing the likelihood of the melted polymer overheating.

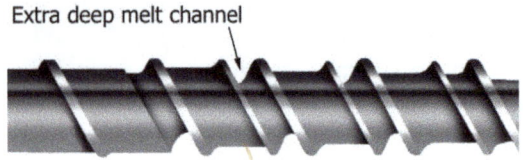

The **plasticizing capacity** is the weight of PS (polystyrene) measured in g/sec that the screw can uniformly plasticize (i.e. raise to a uniform moulding temperature) using maximum screw rotation speed and zero back pressure.

Injection pressure refers to the maximum pressure inside the cylinder during injection, *not* the maximum hydraulic pressure. The two are related by the *intensification ratio* of the screw's cross sectional area to that of the injection cylinder. Injection pressure is considerably higher than the hydraulic pressure since the cross-section of the screw is so much smaller than the area of the injection ram against which the hydraulic pressure is applied. The intensification ratio may be calculated by dividing the maximum injection pressure quoted in the machine specification by the maximum hydraulic pressure the machine is capable of applying to the injection ram.

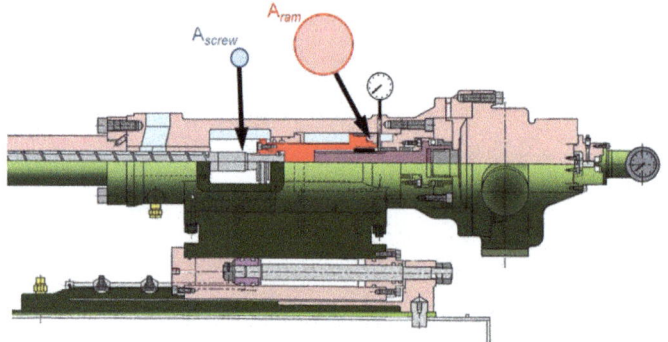

Since the cross sectional area of the injection ram (A_{ram}) is larger than that of the injection screw (A_{screw}), the melt pressure will be proportionately greater than the hydraulic oil pressure.

The **injection stroke** is the *travel* of the injection screw. In practice it is not desirable to use the full stroke as to do so would result in non-homogeneous melt. Although melt consistency ought to be assured by the mixing and shearing that occurs in the metering section of the screw, pellets which enter the feed throat when the screw is fully forward have to travel the full length of the screw, while pellets that enter the feed throat when the screw is towards its rearmost position, travel a shorter distance and so are subjected to less mixing and shearing. As a result machine manufacturer's typically recommend using only between 35% and 65% of the full stroke.

The **injection capacity** is the volume of material that can be shot or injected in one full movement of the screw and is quoted in cm^3. The injection capacity equals the cross sectional area of the screw multiplied by the injection stroke.

$$\text{Injection volume (cm}^3\text{)} = \pi \times (d^2/4) \times \text{stroke}$$

d = diameter of screw [cm]
stroke = injection stroke [cm]

Owing to leakage past the screw tip and the backward movement of the non-return valve, the actual injection volume is typically only about 90% of the theoretical injection volume. To calculate the shot weight in grams, multiply the injection volume by the resin's Specific Gravity measured at the plasticizing temperature.

Resin	Specific gravity*
Polypropylene (PP)	0.712 - 0.737
Polycarbonate (PC)	1.018 - 1.037
Polyethylene Terephthalate (PET)	1.129 - 1.172
Polyethersulfone (PES)	1.240 - 1.580

* at plasticizing temperature

Note that the *residence time* depends upon more than the injection capacity since there is a substantial amount of additional material in the screw flights and much of it is molten. Once this is taken into account, the residence time can be 2½ to 3 times longer than initially assumed.

Sometimes a bigger cylinder than strictly necessary is specified in case a subsequent application should require a larger shot size. This negatively affects the processing of smaller shot sizes in two ways:

- The larger cylinder increases the *residence time* of the raw material in the cylinder. The length of time the plastic remains in the cylinder has a significant effect on the quality of the injection moulding. If it is too short, the pellets will not be sufficiently melted, but if it is too long thermal degradation of the polymer is likely to occur (indicated by discoloration, dark streaks, and even burned particles in the moulded parts).

- For a given injection moulding machine, the shot-size will be increased by selecting a larger cylinder and screw. However on the majority of machines, the diameter of the injection ram doesn't increase when a larger diameter cylinder is selected. This means that as the injection cylinder becomes larger, the intensification ratio translating hydraulic pressure into plastic pressure becomes *smaller*.

The screw rotation speed in RPM (revolutions per minute) is generally quoted because the value can easily be found from the dial at the rear of the cylinder, but in reality the *surface speed* of the screw is more important than the number of times it rotates in a minute:

Screw surface speed [mm/s] =

$$\frac{\pi \times \text{screw diameter [mm]} \times \text{screw rotary speed [rpm]}}{60}$$

Note that for a given RPM, the larger the screw diameter the higher will be the screw surface speed (and thus the shear).

The *counter-pressure* or **back pressure** exerted against the back end of the screw to prevent it from sliding back too readily during charging, helps maintain a constant plasticizing time, minimizes air entrainment (air is carried into the cylinder with the pellets), and improves the homogeneity of the melt (something that is particularly important when additives such as colour are present). Although slight back pressure can be beneficial, too much will lead to excessive shear heating.

The **injection speed** expressed in cm/s is a measure of how fast the injection ram thrusts the screw forward (overcoming as it does so the resistance arising from pushing the viscous melt through the injection nozzle, runner channels, and gate into the cavity).

Some machines are provided with *Injection Controllers* capable of varying the injection speed during injection in order that they may maintain a constant melt front speed even when the cavity cross section varies; a practice that is usually considered to produce the best part quality. In practice, the uniform wall thickness of a typical preform means it is seldom necessary to profile the injection speed.

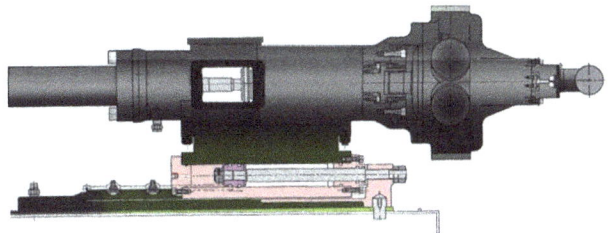

During injection the nozzle on the front of the injection cylinder must be held firmly against the **Sprue Bush** (hot runner manifold inlet) in order to prevent the melt from leaking out, and so the injection unit is mounted on top of an hydraulic actuator. (Note that the piston rod end of this actuator is fixed, and it is actually the cylinder on which the injection unit is mounted that moves).

The reciprocating screw not only has to wait until the mould has closed to make the shot, but is also used to maintain the holding or packing pressure (thus adding to the cycle time, since the screw cannot start charging for the next shot until packing has been completed). With one-stage machines such a delay is unavoidable, but fast cycling injection moulding machines fitted with high cavitation preform tools avoid the problem by having a packing device, colloquially known as a *shooting pot*, located at the front of the cylinder. The screw is used to feed the melt into the shooting pot, then the shooting pot takes over and not only injects the material into the

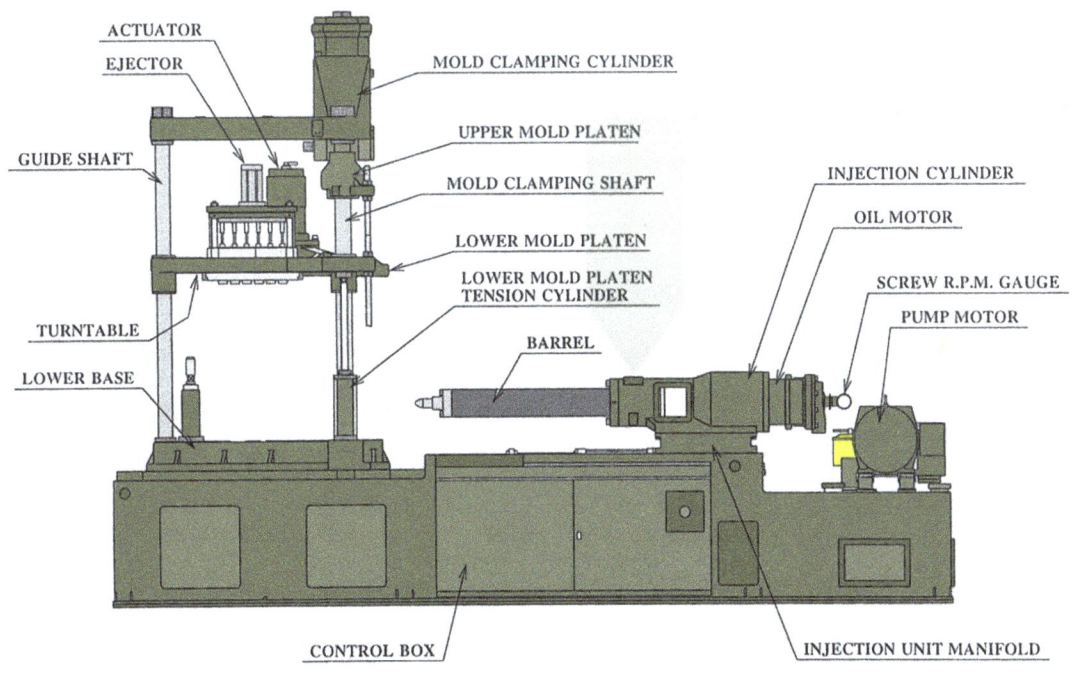

cavities, but also maintains packing pressure on the molten plastic, thus leaving the screw free to plasticize the next shot. Separating the functions in this way allows the extruder to be "sized" for maximum throughput, and the shooting pot for precise shot control.

Heat is generated by both the mechanical action of the screw (*churning* and *compressing* the melt) and by external **heater bands** encircling the injection cylinder. When setting the injection cylinder temperature profile the aim is to achieve the optimal temperature of the melt in front of the screw. A *level* temperature profile is used when maximum heat has to be supplied quickly due to low residence time. A *rising* temperature profile is used when residence times are comparatively long, since it provides fairly mild conditions for melting the material.

Low temperature melt is highly viscous and freezes quickly in the mould (which can result in *short shots* if the gate freezes before the cavity is full), and so higher injection pressures are needed to fill the cavity in a given time. High temperature melt flows more easily, but can result in *flash* or *sink marks*. Sink marks are small depressions in the preform rim caused by inadequate filling of the cavity (the hotter the material the more it will shrink on cooling). Flash is the result of PET flowing into the vents that are provided to allow the displaced air to escape (this happens either because the melt is too fluid, or because too much material is being packed into the cavity and thus forcing it open).

Even though the gate has shut, the mould remains closed and clamped while the final *cooling* of the material takes place. Cooling involves the transfer of heat out of the preform and into the cavity walls by conduction (to be carried away by the mould cooling fluid), and since the plastic is an insulator, thicker parts take longer to lose their heat than do thinner parts. As a result it is the preform wall thickness, rather than its weight, which determines the overall duration of the injection moulding cycle. As the melt cools, it shrinks *away* from the cavity but clings *onto* the core, which makes the injection core pin the most significant component of the injection mould (in so far as the cooling time is concerned).

The actual temperature of the melt can be measured using a handheld probe. Take care when preparing the melt for an *air shot* both to use the regular RPM (because charging the screw with a reduced speed will result in a lower temperature melt), and to pre-heat the probe *above* the anticipated melt temperature, so that it has to cool *down* to the measurement temperature.

In order to stop the operator inadvertently damaging the screw by starting it when cold, a cold-start prevention system is fitted. Less elaborate than it sounds, an alarm "contact" on a temperature

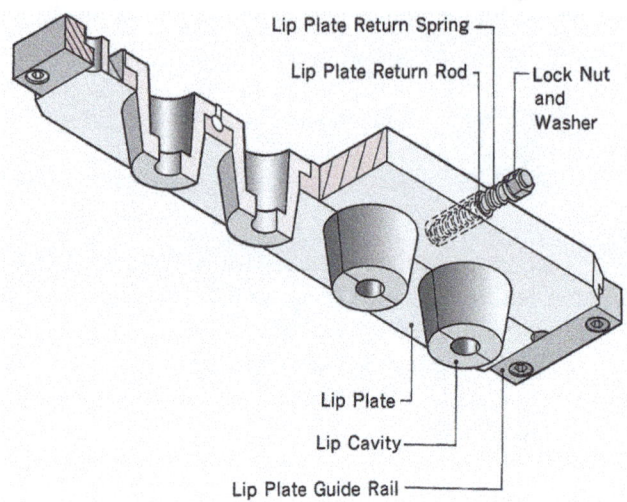

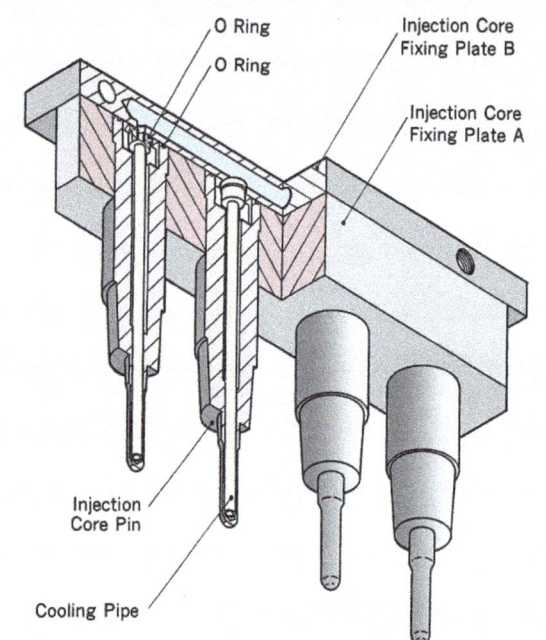

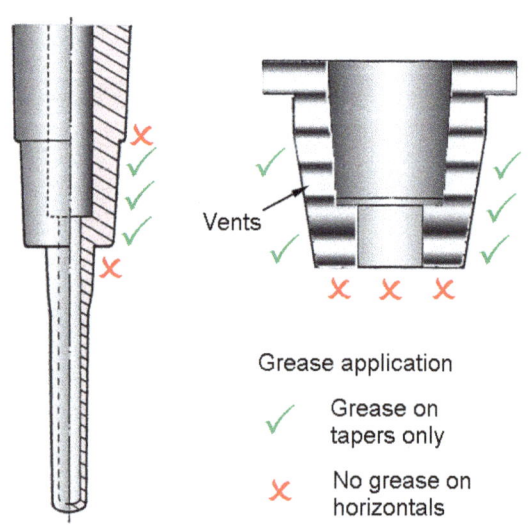

controller simply prevents screw operation when the injection cylinder is below a set temperature. As a secondary protective measure, the holding ring attaching the screw to the injection ram may be secured with "low shear" bolts that are designed to break when subjected to too great a load.

Clamping Unit

The *clamping unit* initially closes the mould, then holds the two halves together during injection and cooling, and finally re-opens the mould at the end of the cycle. The high pressure melt entering the cavities exerts a considerable opening force on the injection mould proportional to the total *projected* area of the part(s) at the parting surface, i.e. the larger the number of cavities, or the bigger the neck diameter, the greater is the opening force (and remember that the injection pressure is calculated by multiplying the hydraulic pressure by the *intensification ratio*). To prevent the mould opening and the melt being squeezed out around the edge (known as *flash*), the clamping force must be greater than the opening force. For PET, the clamping force needed is 3 to 4 tons per square inch of projected part area [420 to 560 kgf/cm2].

The *daylight* is the distance between the fixed and moving platens when the clamp is wide open.

The thickness of the tool must lie within the *shut height range*.

The *mould opening stroke* is the displacement of the moving platen from mould close to mould open. In a conventional injection moulding machine it must be at least twice the height of the moulded part in order that the product can be ejected. In the classic one-stage machine it is only necessary that the mould open sufficiently for the injection core pins to be clear of the rotary table, and for the preform tip not to hit the top of the injection cavity when the table rotates.

Cycle time = Rotation + Mould Close + Clamp Up + Nozzle forward + Injection + Cooling + Mould Open times (assuming blowing has completed).

Injection mould

The classic one-stage machine design is particularly versatile in that with the appropriate toolset installed, the same machine can be used to make a variety of bottles and jars of different shapes and sizes. The toolset includes the injection mould, conditioning parts (on a four station machine), blow mould plus blow core and stretch rod assembly, and the ejectors.

The injection mould, which contains within it the impressions into which the material is injected to give the preform its shape, consists of three parts:

- Paired lip cavities: female portions of the mould which form the outside of the neck.
- The core: the male portion of the mould which forms the internal shape of the preform.
- The cavity: the main female portion of the mould which gives the preform body its external form.

Alignment of the different parts of the injection mould is ensured by matching tapers which both

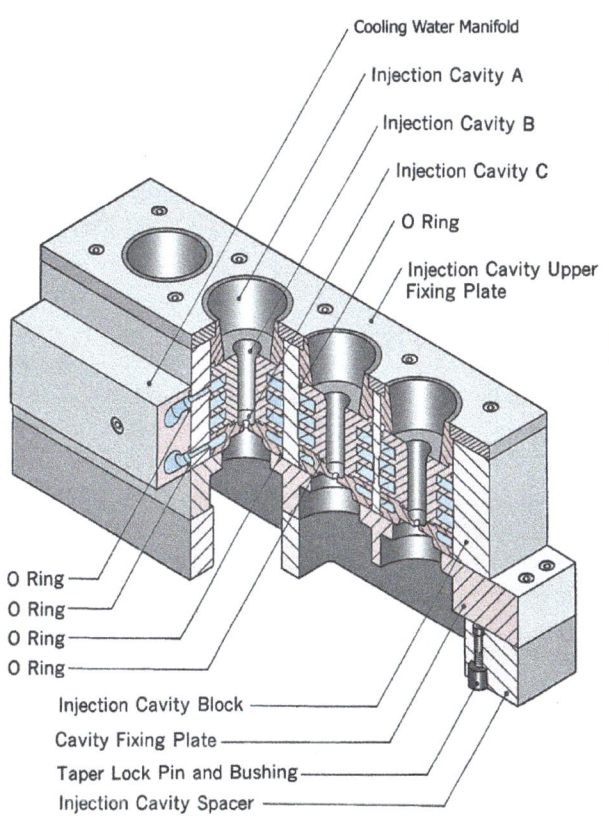

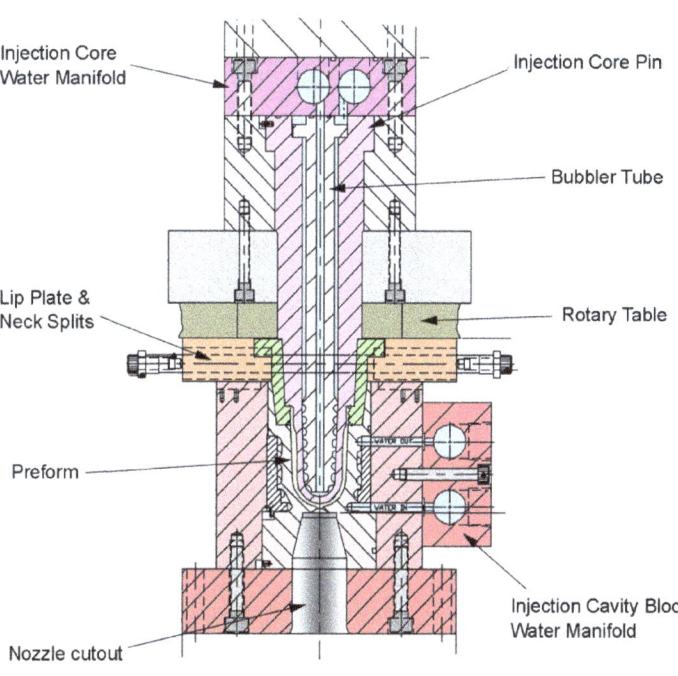

Since PET, like all thermoplastics, contracts as it cools, to compensate the preform injection cavity is typically made oversize by a factor of 1.005.

provide a lead-in to the parts as the mould closes, and a positive location once the mould is closed and held shut by the clamping pressure. If the parts of the mould were not positively located when clamped, during injection the injection core could move, resulting in preforms with non-uniform wall thickness. If eccentricity exceeds 0.1 mm for a preform 120 mm in length (and pro rata for a preform of different length), it needs to be corrected. Even though injection mould parts are typically machined from pre-hardened tool steel, the highly polished moulding surfaces are easily marked; which is why only soft metal tools (e.g. brass) and hammers with plastic heads should be used when working on mould tools.

When the melt is injected into the cavity it displaces the air inside, and so shallow vents (typically 0.03 - 0.04 mm deep) are provided to allow it to escape. If the vents were too shallow (or became blocked with dirt), air trapped inside the mould would prevent it from being completely filled, although equally the vents mustn't be too deep, lest melt flow into them result in *flash* on the preform.

Each injection core and cavity is maintained at a constant temperature by re-circulating fluid through passageways. For effective heat removal, the flow through the "bubbler tube" in the core pin and the channels encircling the cavity must be *turbulent*. The heat extracted initially enters the outer layer of the fluid flowing through the channels, and if the flow were to be *laminar*—meaning the outer layers of the fluid did not mix with the cooler inner layers—the cooling potential would not be fully utilised (in point of fact laminar flow provides only one third the cooling effect of *turbulent* flow).

Turbulent flow is achieved when the *Reynolds Number* is greater than 4000. The Reynolds Number is calculated from the fluid velocity, the diameter of the pipe, and the kinematic viscosity of the fluid. The cooling is most effective when the number is calculated to lie between 4000 and 5500.

Mould cooling not only has a decisive effect on cycle time, it also affects surface quality (gloss) and shrinkage. Higher mould surface temperatures tend to delay solidification of the melt, thereby allowing more time for *packing* and thus greater compensation for shrinkage, but also allow time for crystallization to occur.

Hot runner

During the injection process the PET material is delivered to the mould through a hot runner system. All hot runner systems contain the same basic elements: a *sprue bushing*, i.e. the point where the material enters; *heaters* to maintain the material temperature; *thermocouples* to measure that temperature; and a manifold to distribute the material to each nozzle (through which the molten material flows into the cavity via the *gate*).

The injection cylinder nozzle is pressed against the sprue bushing orifice so that the molten material can flow directly into the manifold. To withstand the high loads, the sprue bushing is a replaceable hardened steel insert. It has a curved contact surface with a slightly larger radius than that of the spherical nozzle tip to provide a good seal without creating a material hang-up point.

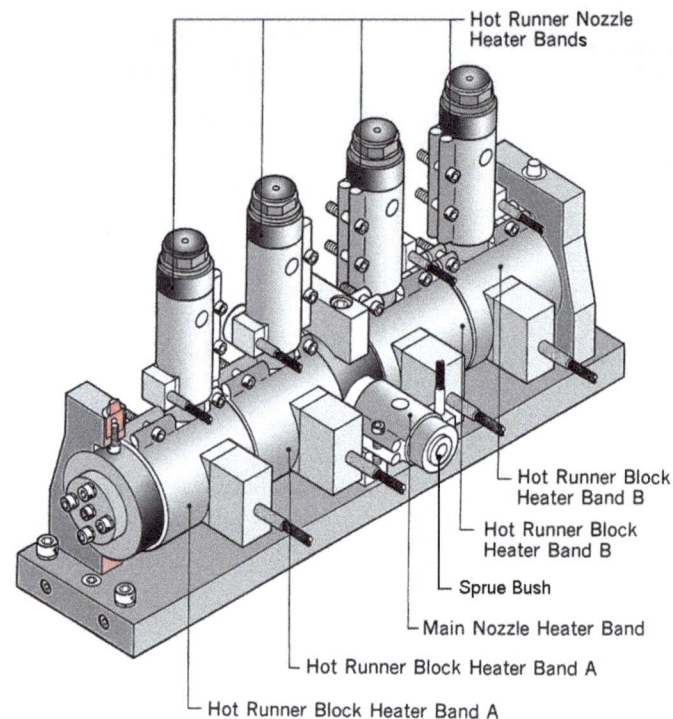

The manifold conveys the melt to the individual nozzles through a system of channels known as *runners*. Determining the appropriate runner diameters involves a compromise between the need to avoid excessive pressure drops and minimize shear heating effects, simultaneously allowing for the fact that high viscosity resins require larger diameter runners than do low viscosity resins, while also minimizing the time for which the material resides in the hot runner system! Manufacturers take pains to give runners a fine surface finish that will ensure a smooth flow path with no material hang-up points which might otherwise cause the polymer to degrade inside the hot runner system.

A uniform flow of material into each cavity is essential since the material starts being cooled as soon as it enters the cavity (meaning that a preform from a cavity which had filled very rapidly would be colder than one from a cavity that had filled less quickly). Such a preform temperature difference would cause cavity-to-cavity variation in the blown containers. One of two different methods may be used to provide the required *balanced* fill:

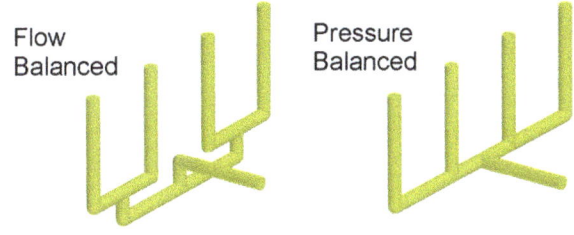

- Ideally the manifold will have equal flow lengths to each cavity (known as a *flow balanced* manifold), although this can make it more costly to manufacture.
- With the alternative *pressure balanced* system the temperature of each nozzle is independently adjusted in order to control the flow of material by varying its viscosity, while the nozzle orifice diameters are individually sized to deliver the desired flow of material.

The hot runner *nozzle* is the connection between the manifold and the gate, and must be capable of shutting off the flow when required. A valve gated system utilises a *valve pin* to provide a positive break, whereas a thermal-break system depends upon the melt in the gate orifice freezing and plugging the gate while the holding pressure is being applied. During the next cycle, injection pressure forces the solidified material in the gate orifice into the cavity where it fuses with the melt. When a thermal break is used, a physical shut-off is also required on the nozzle fitted on the front of the injection cylinder so that the cylinder may be charged without either pressurising the hot runner manifold, or drooling if it is retracted.

Open nozzles have some advantages: they are simple and reliable, and don't result in flow lines.

The gate orifice diameter is generally smaller than with a valve gate to ensure that the material in the gate region doesn't take too long to solidify (which leads to *stringing*), although in order to prevent excessive sheer heating, the diameter of the nozzle aperture should not be less than 2.8mm. The most visible disadvantage of the thermal-break nozzle is the length of the gate remnant, although modern nozzle tip designs ensure this is not more than 2-3 mm in length.

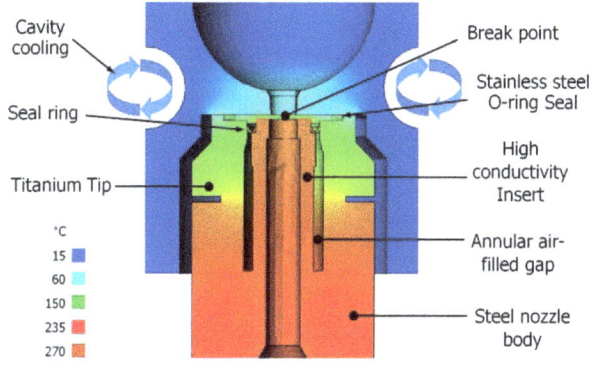

Computer-generated cross section of a thermal break nozzle showing typical temperature gradient

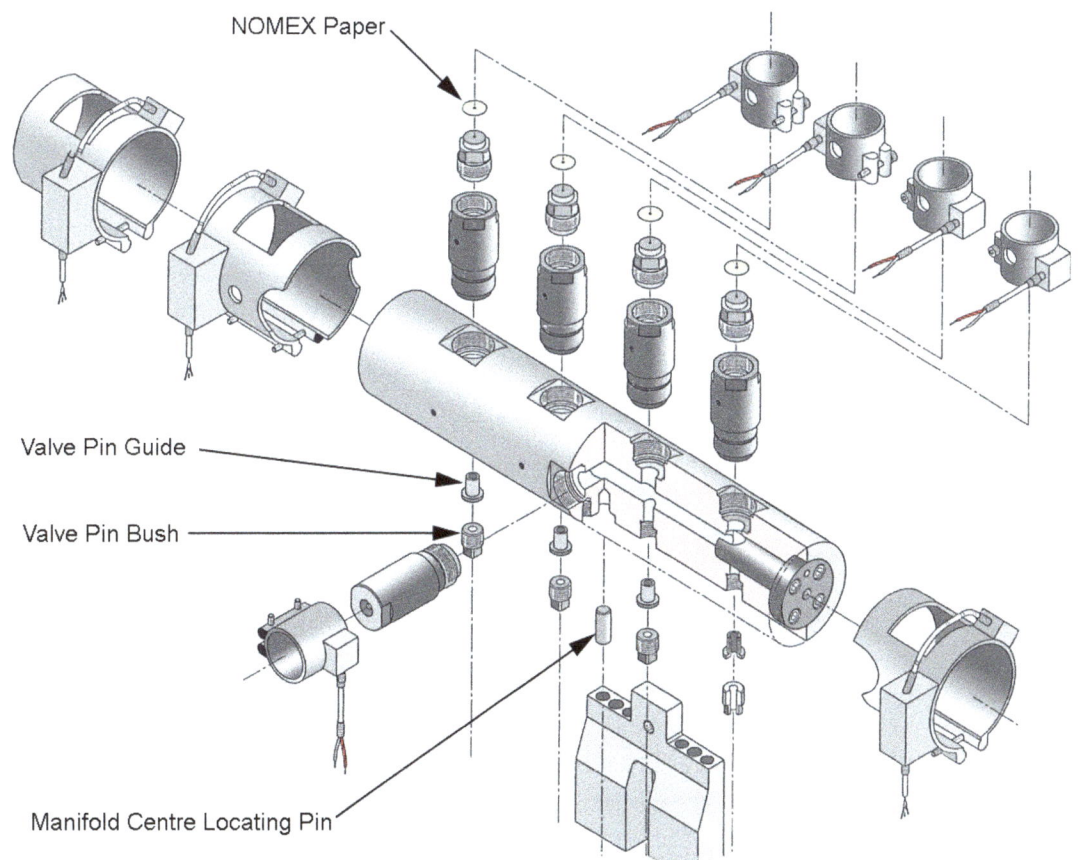

Unfortunately thermal-break gate quality can change significantly as processing conditions vary. For example *Mould Open* cannot begin until the gate is solid enough to break cleanly from the part (lest drool or stringing occur), while *melt decompression* is required to relieve pressure inside the manifold (even though this can sometimes lead to splay marks and other visual imperfections). Cooling of the cavity is more critical in thermal-break systems: inadequate cooling in the gate region will mean that the holding pressure has to be maintained for longer — adding unnecessary seconds to cycle time; whereas excessive gate cooling can cause frozen gates resulting in short shots or unfilled cavities.

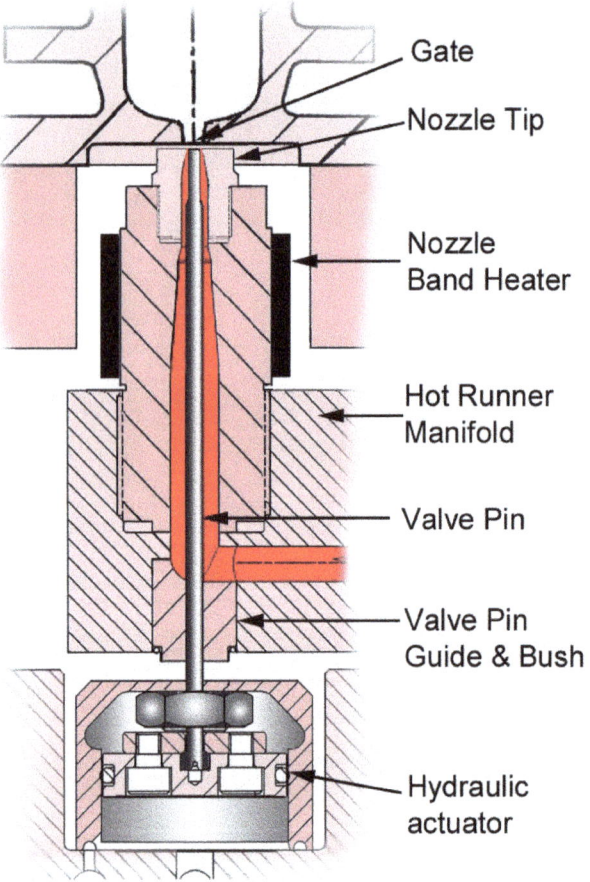

A valve gate nozzle uses a valve pin to positively seal the gate orifice, thereby eliminating drool and stringing. The pins may be pneumatically or hydraulically actuated. Aoki machines use the latter, meaning that when hydraulic pressure is applied to the injection ram to force the screw forward and inject the material, hydraulic pressure is simultaneously applied to the actuators mounted underneath the hot runner manifold that open the valve pins. Similarly after injection and packing have been completed, hydraulic pressure closes the valve pins so that plasticization can start immediately (thus minimising the cycle time).

Valve gates offer more consistent product than thermal gates because small changes in the processing parameters or fluctuations in the melt viscosity do not affect their performance. Disadvantages include the likely appearance of weld lines from where the melt has flowed around the pin, and the necessity to regularly replace worn valve pins. Generally however these shortcomings are outweighed by their advantages:

- consistency cavity to cavity and shot to shot
- more time available for plasticization since it can begin as soon as the valve gate closes, without having to wait for the melt in the gate orifice to freeze
- Minimum gate vestige for better aesthetics
- Large gate sizes induce less shear heating
- Elimination of drool during plasticisation, making injection cylinder shut-off unnecessary

Pin ends may be conical or cylindrical, and their length is critical. If too short they will not seal effectively, leaving a poor gate vestige and drooling from the gate. If pins are too long, as they warm up and the steel expands, they will damage the gate pad. For this reason a cylindrical shut-off pin (if necessary provided with a conical shoulder to pre-guide it towards the centre of the gate), is usually considered preferable.

Calculation of Pin thermal expansion:

$$E_{Pin} = L_{Pin} \times \alpha \times \Delta T$$

In order to ensure a crisp, clean gate when the part is ejected, the temperature of the front end of the valve pin needs be less than the melting point of the polymer, which means nozzles should be operated at the minimum temperature that enables satisfactory operation. Note that when hot runner systems are cold, nozzle centre-to-centre distances are undersize. Manifolds are pinned to the base plate at their midpoint and as they heat up, allowed to expand outward from the centre (dowel pins sliding in slots maintain the correct orientation). Cooling water flowing in the manifold base plate prevents heat transfer to the valve pin actuators or lower movable platen (thermally induced expansion would cause alignment problems).

Description	Variable	Units	Value
Expansion	E_{Pin}	mm	
Pin length in cold condition	L_{Pin}	mm	
Nozzle Temperature	T_{Nozzle}	°C	
Mould Temperature	T_{Mould}	°C	
Change in Temperature	ΔT	°C	$T_{Nozzle} - T_{Mould}$
Coefficient of Thermal Expansion	α	1/°C	0.0000132

Most hot runner systems are very slightly taller than the "stack" into which they fit so that a pre-load condition exists on assembly. Thermal expansion of the metal will further increase the compression on the system at operating temperatures. Aoki hot runner systems are different, in that a clearance is maintained between the nozzle tip and the gate pad into which an insulating *Nomex* paper is placed. The paper helps to thermally separate the hot manifold from the cooler mould while also sealing between the nozzle tip and gate. The high temperatures and clamping forces to which they are exposed mean Nomex papers don't survive very long and therefore need replacing on a regular basis.

A molybdenum-based anti-seize compound should be applied to *external* threads when re-installing nozzles and tips.

Conditioning

To achieve the desired material distribution in a container requires both the right preform design, and the correct preform temperature profile, since the hotter parts of a preform stretch more easily than the cooler parts. If the preform temperature is too high, the preform will blow unevenly because the designed level of orientation won't occur and the bottle may even be hazy in appearance due to crystallinity. Whereas if it is too low, microscopic tears in the PET material will form on the inside surface of the container resulting in a pearlescent appearance (assuming the preform doesn't burst inside the blow cavity).

Ideally the preform will have the correct temperature profile when extracted from the mould (or very soon after), but as we have seen, the use of alternative grades of material, changes in ambient conditions, and particularly process changes, can all influence the preform temperature profile. Therefore on four-station machines the preform temperature profile can be adjusted at the conditioning station. Since different bottle designs and performance specifications require different material distributions in the finished container, a range of conditioning methods have been developed to provide the appropriate degree of temperature adjustment for each application:

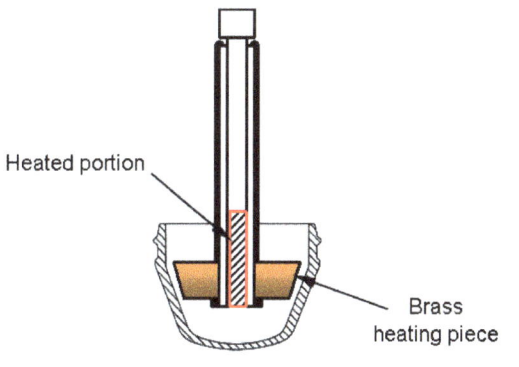

An internal *Heating Piece* may be used to counteract the cooling of the PET immediately adjacent to the neck splits (lip cavities). This consists of an electrical resistance heater heated only at the lower end, and fitted with a brass or copper disc-shaped heating piece that fits inside the preform but without touching it.

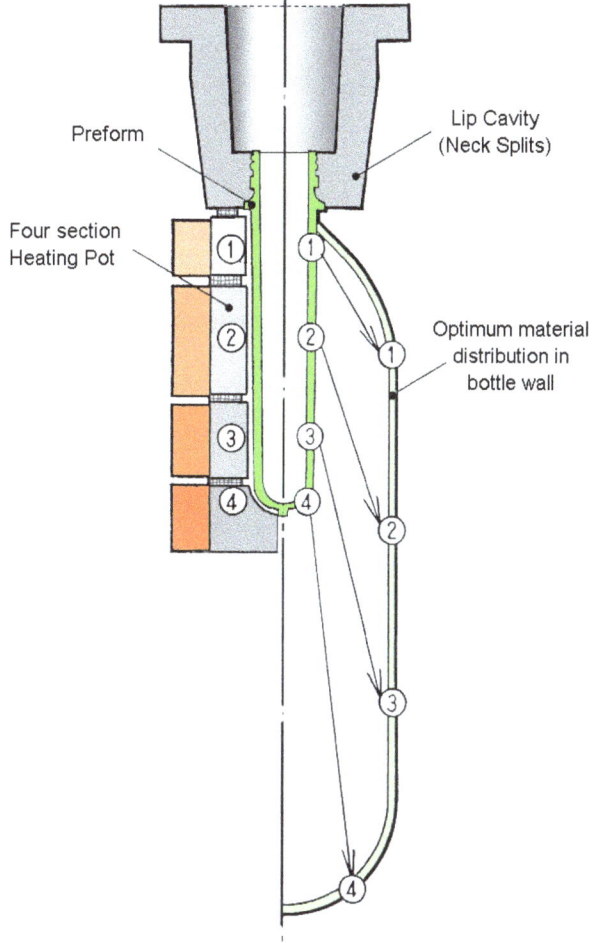

An external *Heating Pot* may be used to warm the outside of the preform without touching it. A heating pot consists of a stack of one, two, or three aluminium rings plus an aluminium cup at the bottom of the stack, all thermally isolated from each other, and with individual heater bands and thermocouples. Each segment may be individually contoured to direct the heat precisely where it is needed (note that the preform doesn't touch the heating pot, but is warmed purely by radiated heat).

With a *Conditioning Core*, hot thermo-oil or water from an external Temperature Controller circulates through a polished aluminium core which touches the inside of the preform (except for any areas where the metal of the core has been cut away). Typically the temperature of the oil or water will be around 70°C, and so the conditioning core actually cools the preform, which for PET will likely still have a temperature of 110-120°C.

Conditioning Blow can be used when stretch-blow moulding materials such as polypropylene (PP) that don't appreciably increase in strength when stretched. Without the automatic wall thickness control imparted to PET by biaxial orientation, attaining an even wall thickness in a PP container depends utterly on the accuracy of the preform mould and the uniformity of the preform heat content. What makes this difficult to achieve is the fact that PP has a lower density and is less rigid than PET, and so in order to achieve adequate handling performance, PP bottles—and thus of course their preforms—must be made thicker; but because thick wall sections retain the heat for longer, they contract significantly as they cool, leaving areas of the preform where the hot plastic has shrunk away from the moulding surface hotter than parts where shrinkage has been less. In such cases it is not sufficient to depend upon the warming effect of the heat radiating from a heating pot, but the preform must actually touch the inside of the heating pot (which is therefore polished to a mirror-like surface finish). To avoid creating scratches on the outside of the preform, once the heating pot is in position, low-pressure *Conditioning Blow* air very slightly expands the preform until it presses against the inside of the larger pot, (but not so much that the preform won't shrink back elastically when the pressure is released). To allow for the possibility of cooling as well as heating the preform, the pot temperature is adjusted by the circulation of thermo-oil from an external Temperature Controller.

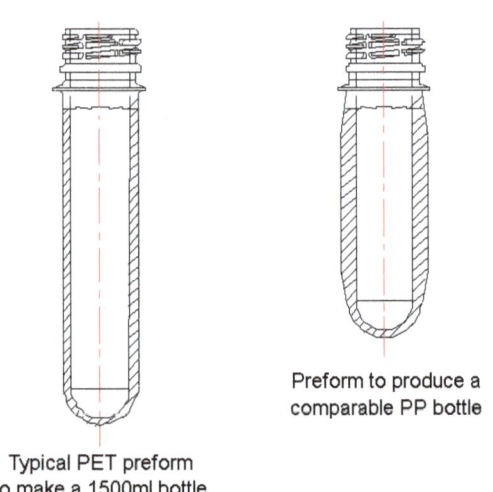

Typical PET preform to make a 1500ml bottle

Preform to produce a comparable PP bottle

When the PP is distributed over the large conical surface of a wide-mouth container (jar) preform, its thickness may not be a concern, but in a tubular bottle preform, PP's low density means the preform wall is extremely thick. In such a case the *Conditioning Blow*-induced slight expansion of the preform inside the heating pot won't thin the preform wall sufficiently to enable rapid heat transfer, so the preform has to be inflated and expanded much more significantly. In this case the conditioning blow will stretch the PP well beyond its elastic limit, meaning that when the air is released the preform doesn't shrink back to its original size, and so a split-mould type heating pot is required.

Stretch-blow moulding

At the blowing station the hot preforms are stretched and blown into blow moulds using high pressure blow air (up to 2.5 MPa). The stretch-blow process uses two different pressures to blow the container: low pressure referred to as P1, and high pressure known as P2 (they are also called primary and secondary). To ensure the flow of blow air is not restricted (and also to minimise any pressure fluctuations arising from changes in the load or in the supply), the blow air is taken from a high pressure Air Receiver fitted inside the machine near to the blowing station. This Air Receiver has a Safety Relief Valve to guard against over-pressurisation, a Pressure Gauge, and a manual drain valve.

The P1 air is drawn from the Air Receiver through a high flow rate pressure regulator like the "dome" regulator pictured right. The control port of the dome regulator is *charged* by a spring-loaded manually-operated pressure regulator (such as the one shown below

right). This "Remote Control Valve" is then used to pressurise the dome to the required P1 pressure. Using air rather than a spring to set the P1 pressure enables it to be adjusted remotely, meaning that setting changes can be made much more rapidly and accurately.

The P2 pressure is simply the Air Receiver pressure, and in order to achieve the maximum preform inflation rate, the air is fed via a manifold to each Blow Core through a dedicated hose.

The stretch rod not only keeps the preform gate centred in the base of the container, but together with the primary blow air, helps *place* the material correctly in the blow mould. The precise times in the stretching and blowing sequence when the primary and secondary blow are initiated is critical since this—together with the temperature of the preform—controls the material distribution in the container. For example, the amount of material in the lower half of a container can be increased at the expense of that in the upper half, by delaying the start of the primary blow to allow the stretch rod more time to drag material towards the base of the bottle before the blow air begins to expand the preform sideways. Once the expanding preform has touched the blow mould, the material hardens and no more stretching is possible.

In comparison to the extrusion blow-moulding process, much less blow mould cooling is required when blowing PET, due to its lower blowing temperature (~100°C). Blow moulds are cooled (or heated when moulding polymers such as polycarbonate and polyethersulfone) by circulating either with water or oil (the latter is employed when temperatures above 95°C are required) through channels in the blow mould back plate.

The pressure experienced inside a blow mould cavity is significantly lower than that inside an injection mould (~2.5 MPa compared to tens of MPa), meaning that much lower clamping forces are involved, and typically aluminium rather than steel is used for its construction. Although blow air pressures are a fraction of those experienced in injection moulding, they are still high enough for any blow mould surface imperfections to be reproduced on the bottle, and so the mould surfaces must have no visible defects, as well as being highly polished. This is even more important when using higher blow mould surface temperatures, because any defects will be reproduced more readily.

The volume of blowing air that will be needed is calculated by adding the capacity of the blow air manifold, hoses, and blow mould(s), and multiplying this both by the number of cycles per hour, and by the supply pressure to the machine (2.5 MPa). The totals are expressed in m^3/hr FAD (Free Air Delivered). An example calculation can be seen in *Chapter 8—Plant engineering*.

Blowing air needs to be oil-free to avoid any possibility of oil contaminating the containers. Oil-free compressed air is produced by either using an oil-free compressor (oil-free bearings, cylinder liners, and low friction heat resistant PTFE piston rings enable the compressor to operate without the use of lubricating oil). Or by using a lubricated compressor in tandem with dedicated oil separation and filtration equipment. Oil-free compressors usually have higher capital costs, lower efficiency, and higher maintenance costs than oil-injected compressors. One disadvantage of using a lubricated compressor is that the need for additional compressed air separation and filtration equipment does reduce the overall efficiency of such systems (particularly if they're not properly maintained). When deciding between an oil-free or a lubricated blow air compressor, the capital cost of the equipment should be weighed against the potential cost to the business of containers being accidently contaminated with oil. Issues to consider range from internal (possible tainting of the contents), to external (poor ink adhesion on greasy containers).

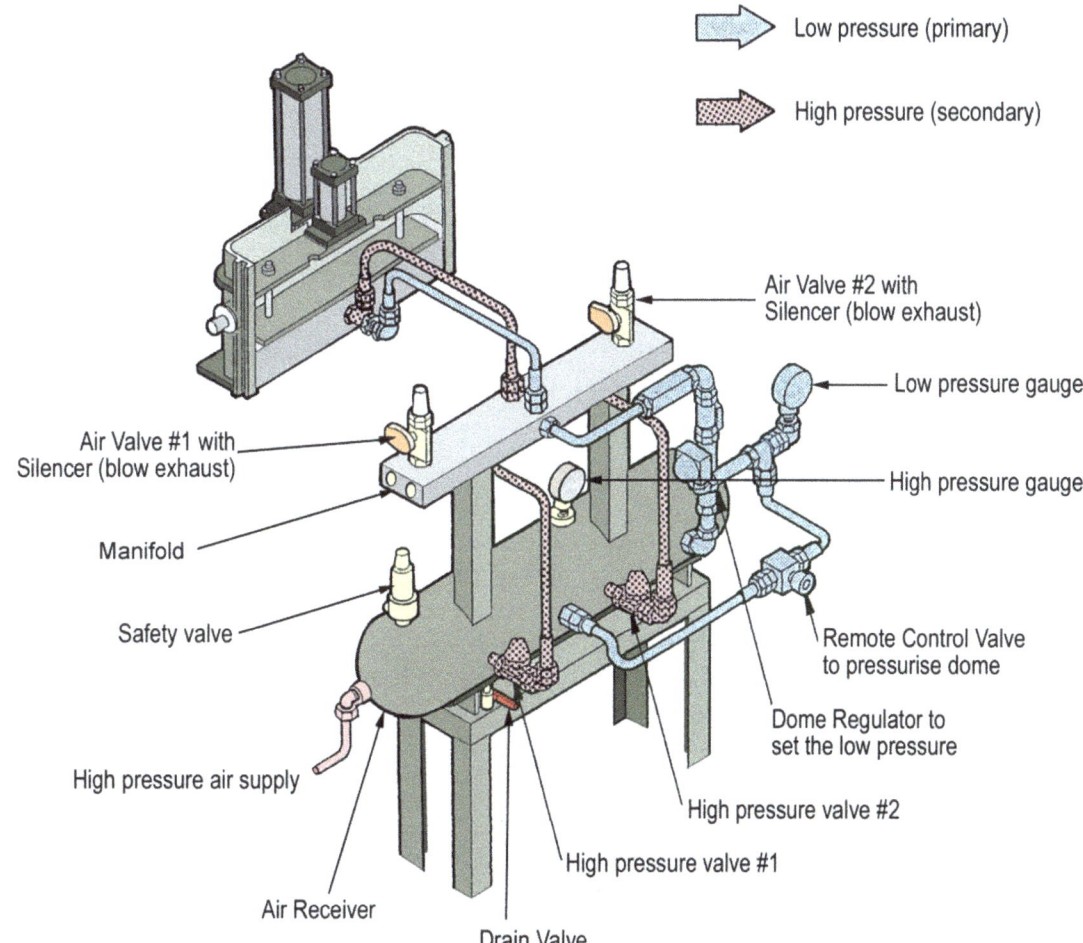

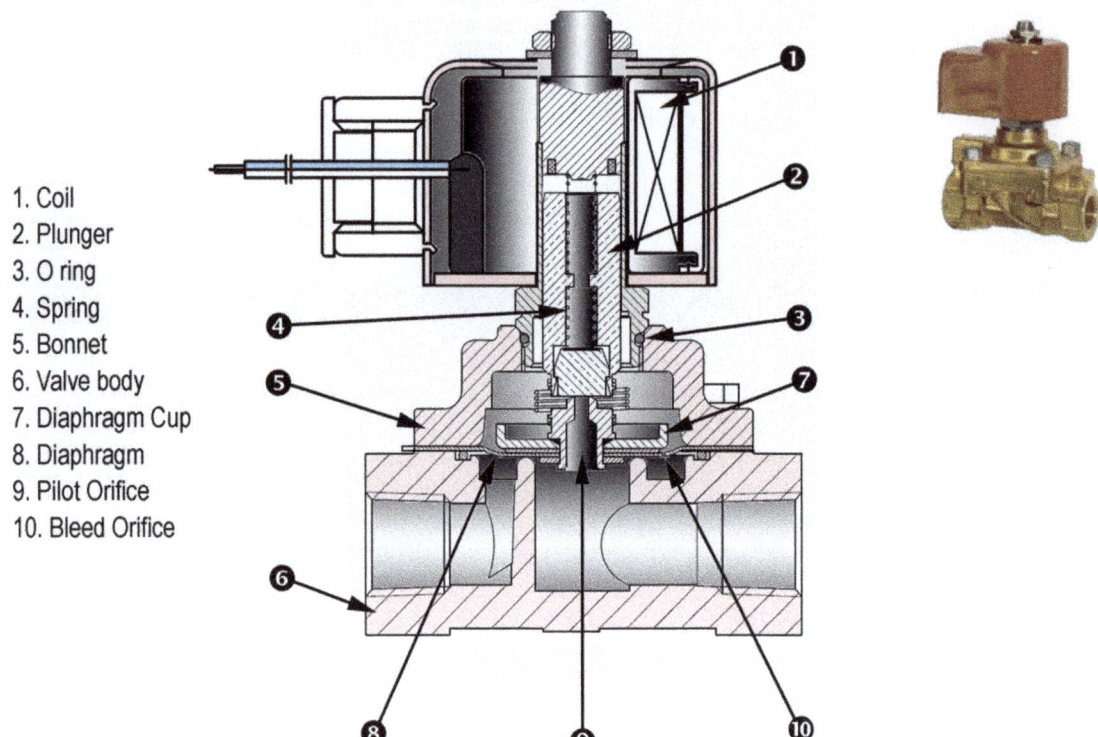

1. Coil
2. Plunger
3. O ring
4. Spring
5. Bonnet
6. Valve body
7. Diaphragm Cup
8. Diaphragm
9. Pilot Orifice
10. Bleed Orifice

Pilot Operated 2-Way Valves

The P1 and the P2 blow air are unleashed by high flow rate normally-closed solenoid-operated 2-way valves. Unlike with hydraulic systems, the blow air system does not use spool valves. This is because very large spools would be needed to deliver the high flow rates necessary for rapid inflation, but large spools don't open quickly enough and so fast-acting diaphragm-operated solenoid valves are used.

Solenoid valves are electrically operated devices used to control flow. They are used for the remote on/off or directional control of fluids. They do not regulate flow. Solenoid valves consist of two main elements: 1) an electrical coil in the solenoid, and 2) a valve body. The solenoid is the electromagnetic unit that powers (acts to open or close) the valve. The valve is the pressure containing unit that acts to shut off or open the air flow.

In general, solenoid valves are constructed to be either Normally-Open, or Normally-Closed. Both designations refer to the action of the valve on flow when the solenoid is not energized. There would be, for example, no flow through a normally closed valve until its solenoid was energized.

The most common types of solenoid actuated valves are: 1) Direct-Acting, and 2) Pilot-Operated. In a direct-acting valve, the plunger is in direct contact with the body main orifice, and opens or closes the orifice. In a pilot-operated valve, the main orifice is not directly controlled by the plunger, but by a pneumatically operated diaphragm or piston. Blow air systems use internally pilot-operated valves because direct-acting valves would require very large power-hungry solenoids to power them (due to the high flow rates and pressures).

Three 2-way solenoid valves control the primary and secondary blowing and the exhaust functions:

- The NO Exhaust1 valve(s) and NC Primary blow air2 valve(s) are energised so that the pressure in the circuit rises rapidly to the pre-set level of 0.2 to 25 bar.
- The NC Secondary blow air3 valve is energised so that the pressure in the circuit increases up to 25 bar. The design of the Primary blow air2 valve ensures that the 25 bar pressure is maintained and cannot go back into the low pressure circuit.
- The NC Primary blow2 and NC Secondary air3 valve(s) are de-energised.
- The NO Exhaust1 valve is de-energised to permit the blow air to discharge through the silencer.

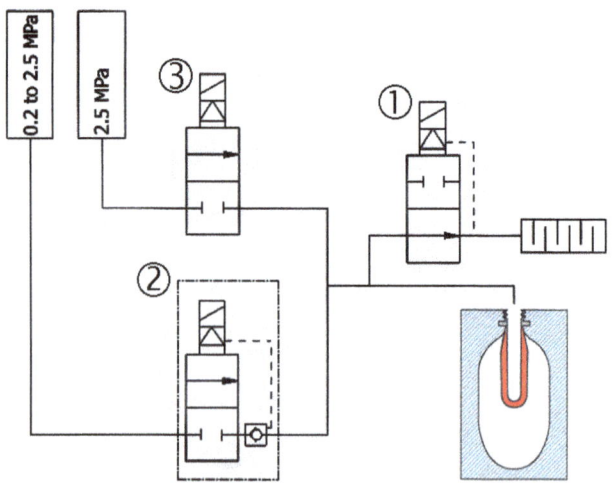

Operational Sequence for a pilot-operated, Normally Closed 2-way valve

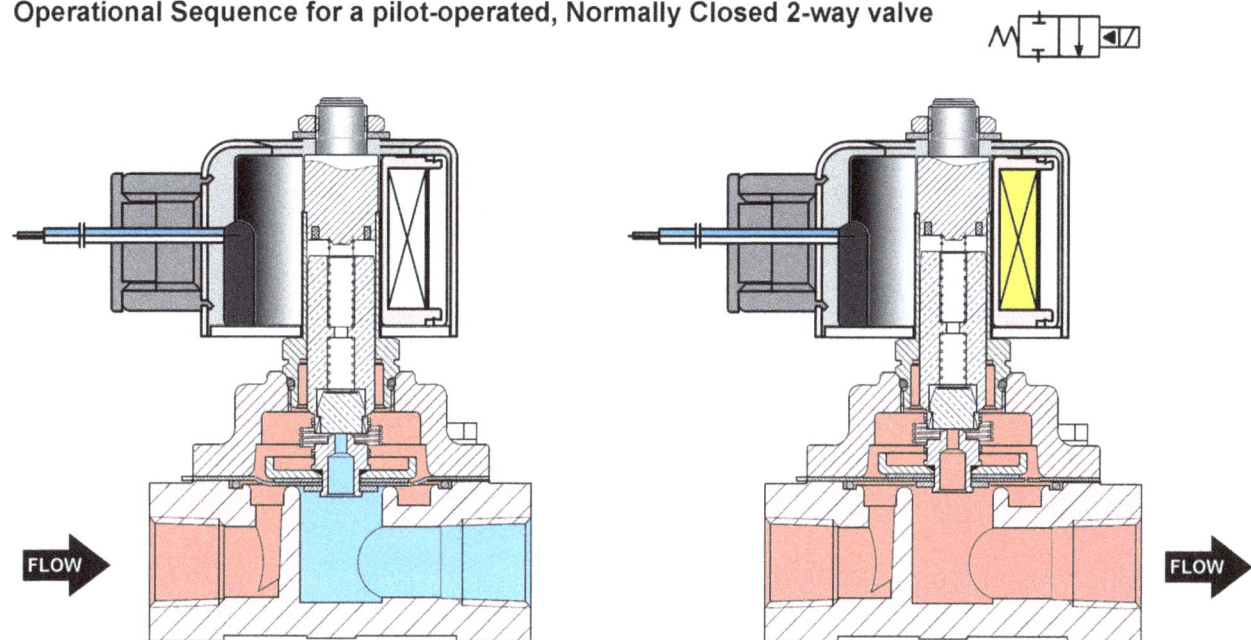

Normally Closed, De-Energised

Closed: When the solenoid is *de-energised*, it no longer holds up the plunger, which is pushed down by a spring. When the plunger drops, a disc at its lower end covers and closes the pilot orifice through the centre of the diaphragm. Compressed air continues to pass through a bleed orifice in the diaphragm until the pressure has equalized in the upper chamber. Since this pressure is acting against a larger surface area on the top face of the diaphragm than on the bottom, it results in a net force downwards which holds the diaphragm firmly against its seat, thus stopping any flow through the valve.

Normally Closed, Energised

To Open: When the solenoid is *energised*, a magnetic field is formed which attracts and lifts the plunger. The disc no longer covers the pilot orifice through the centre of the diaphragm, and so the pressure in the upper chamber bleeds away through the now open pilot orifice.

As pressure on top of the diaphragm is reduced, the pressure acting on its underside lifts it off the seat, thus allowing full media flow through the valve. Since the bleed orifice has a smaller diameter than the pilot orifice through the centre of the diaphragm, system pressure cannot rebuild on top of the diaphragm and close the valve, as long as the pilot orifice remains open.

	Pressure equivalence table						
psi	5	10	14.5	145	290	362	435
bar	0.3	0.7	1	10	20	25	30
MPa	0.03	0.07	0.1	1	2	2.5	3

28   Chapter 2—Injection stretch-blow moulding

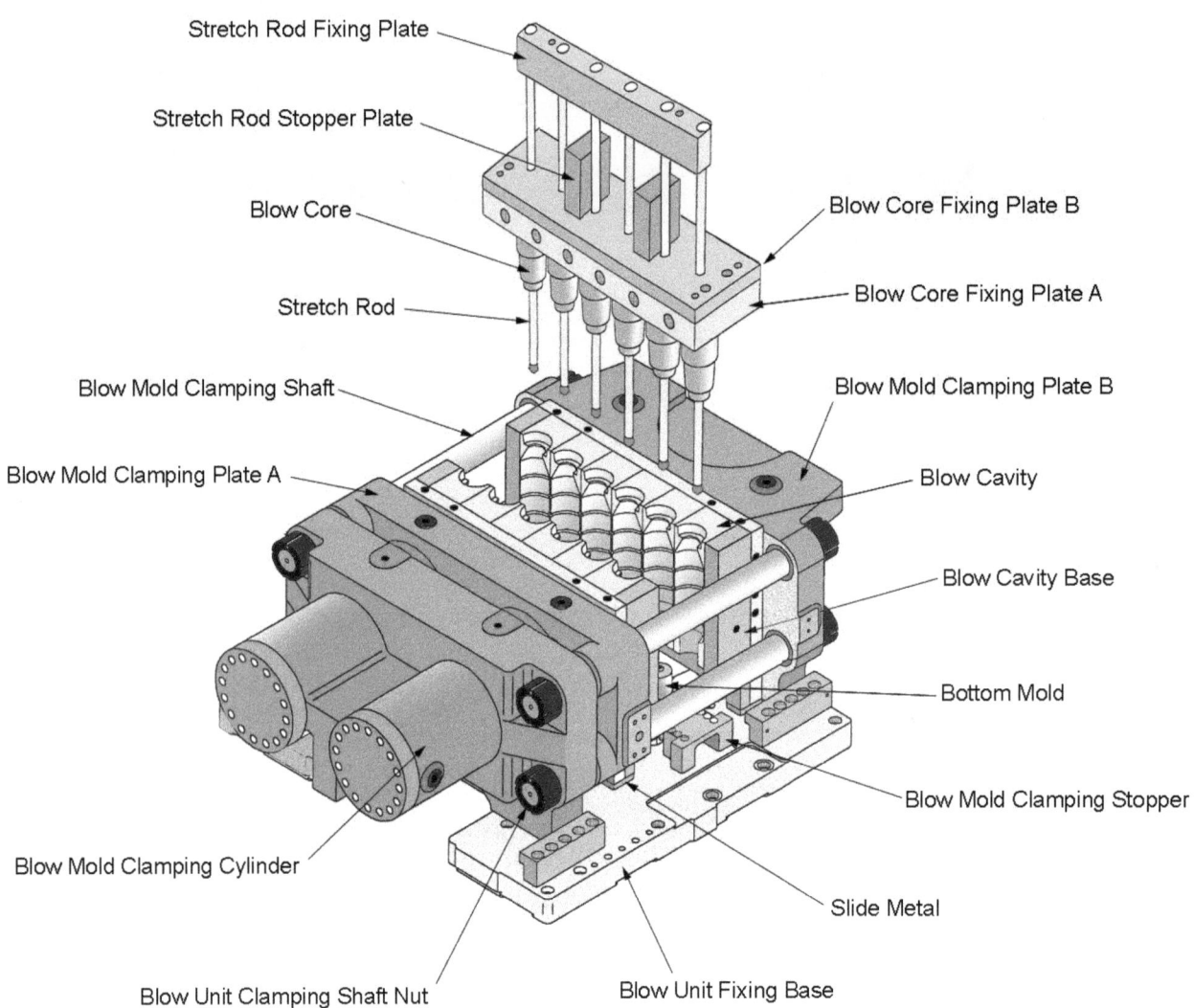

3

Materials

Classic one-stage machines are now used to injection stretch-blow mould a wide range of container designs in a variety of materials, but they were originally designed to produce poly(ethylene terephthalate) bottles, which means it is the physical properties of PET that have defined the machine layout. Therefore this chapter focuses mainly on this material, although alternatives such as PP (polypropylene) are also described.

The first question to address has to be what is it about plastic materials that makes them so useful? While they are not as strong as most metals and are generally unable to withstand very high temperatures, these disadvantages are outweighed by plastic's many positive attributes including being durable, lightweight, self-coloured and in some cases transparent.

Assuming it was possible to examine the internal structure of a plastic, it would be seen to be made of *macromolecules* consisting of many small identical units linked together in a chain (which is why such materials are referred to as *polymers,* from the Greek for "many parts"). Polymers may conveniently be divided into two categories: *thermosetting* and *thermoplastic*. The macromolecules from which thermosetting polymers are made are heavily *cross linked*, giving a rigid structure. Once formed, thermosetting polymers do not soften when heated. In contrast thermoplastics such as PET have few cross linkages, so that when they are heated they gradually soften until they become viscous liquids, then when they are cooled their original properties return. As a result a variety of forming techniques—including injection stretch-blow moulding—have been devised to economically produce as many parts, often with complicated shapes, as may be needed.

The origin of the "mould-ability" of a thermoplastic such as PET is its gradual softening from a solid-like material to a liquid-like material when heated. To melt a polymer requires the input of sufficient heat energy for the macromolecules to be able to move relative to one another. However when they start to absorb this heat, some parts of the molecule will receive sufficient energy to move before than other parts; meaning that the material as a whole is neither solid nor liquid, but *plastic*, and thus can be moulded.

Only thermoplastics may be injection moulded or blow moulded, but not every thermoplastic is suitable for injection stretch-blow moulding.

LLDPE macromolecules

Atomic Force Microscopy
Scan size 8μm x 8μm)

PET: Facts & Figures

PET is a thermoplastic polyester manufactured in two stages:
- the monomer created by combining terephthalic acid and ethylene glycol is condensation polymerized in a melt phase reaction to form a relatively low IV amorphous PET.
- this is subjected to solid phase polymerization to give a higher IV crystalline PET.

Key properties include:
- Intrinsic Viscosity (IV), a number correlating to the average length of the molecular chains from which the polymer is formed. It determines not only the viscosity of the molten polymer, but also the toughness and impact strength of the bottle
- Glass Transition Temperature (T_g) which determines the temperature at which the amorphous polymer will soften
- PET can be crystalline or amorphous according to the heat treatment it has received. Amorphous material is transparent with a high gloss, whereas crystalline material is opaque white

There are two ways to make PET crystallize:
- by heat, which results in a white brittle material
- by mechanical stretching of the PET to orientate the molecules which gives a material *with improved physical and barrier properties*.

Acetaldehyde (AA), which can cause an off-flavour in mineral waters and cola drinks, is generated in small amounts during the melt processing of PET. It can be minimized by using the mildest possible moulding conditions.

Specially modified grades of PET known as *co-polymers* have been developed to provide specific material properties such as slower rates of crystallization.

PET is *hygroscopic* meaning it absorbs water from the atmosphere. This must be removed before processing since its presence in the PET melt will cause the polymer chains to break down, (i.e. reduce the IV and associated physical properties).

To achieve the maximum strength improvement through orientation, PET must be stretched by the correct amount (known as the *natural stretch ratio*), and at the correct temperature. Preform and bottle designs, material selection and process settings must all be made with this in mind.

In order to be capable of injection stretch blow moulding, the polymer must possess a temperature range over which it is soft enough to be blown, while retaining sufficient cohesion not to burst in the process. If the melt strength of the material were to be too low, the preform would sag uncontrollably, making it impossible to blow consistent bottles. Whereas if the cohesive strength were to be too high, the preform would simply shatter (assuming that is, the blowing pressure was strong enough).

Crystalline or amorphous

How quickly a polymer is cooled from the melt can make a great deal of difference to the underlying molecular arrangement and thus its properties. PET is a classic example which crystallizes if it is cooled slowly, but when cooled rapidly, retains the molecular organisation of the liquid; i.e. becomes a clear *amorphous* (unstructured) glass-like solid.

Many solid materials are crystalline, where the atoms are arranged in regular, repeating patterns. Glasses in contrast have no underlying regular structure; instead, their atoms or molecules are jumbled together. They may be packed in so tightly they cannot move, but they are not packed in a regular way. What's bizarre is that structurally, glasses are the same as liquids—when looking at a microscopic image of the atoms or molecules, it's not possible to tell the difference between the molten and glass states. When molten glass is cooled, it flows more and more slowly (its viscosity increases), until at some point it's essentially solid. Unlike water, which is either water or ice, glass smoothly changes from a fluid to a really slow fluid, to a glass.

So why is a glass a glass, and not a liquid? Picture the molecules as being free-moving particles, as they cool down the particles pack together so that if one of them wants to move, its neighbours have to cooperate. As they pack ever more tightly, more of the particles have to cooperate for any to move. When all of the particles in the sample have to cooperate, the sample is essentially a solid.

When any polymer is molten, its constituent macromolecules tend to be tangled indiscriminately as in a bowl of spaghetti. If it is then cooled very quickly (e.g. as happens when a preform is injection moulded), the macromolecules will be frozen in their haphazard state, and the polymer is said to be amorphous (and in the case of PET, transparent). However if the constituent macromolecules happen to be made up of monomers that are very regular in shape[2], and cooling is not instantaneous, but allows enough time for them to be able to move into a

---
[2] Unlike in polystyrene for example, where phenyl groups (C_6H_6 rings) are randomly distributed on both sides of the polymer chain, thus preventing them from ever aligning with sufficient regularity to form crystals.

regular close-packed lattice-like array, the polymer will *crystallize*. The longer the chains, the longer it takes for the material to crystallize.[3] For PET the peak crystallization rate occurs at around 165°C, and the heat induced crystals, despite their small size, are large enough to disrupt the transmission of light, meaning that crystalline PET is white. (In point of fact, because its macromolecules are so "long", PET never fully crystallizes but consists of crystals in an amorphous matrix).

The degree of crystallinity affects the way in which the polymer melts: when a clear amorphous sample of PET (i.e. with negligible crystallinity) is heated, it first begins to behave like rubber. The phase change from a solid to a rubbery material is termed the glass-rubber transition, and the central temperature of this range is referred to as the glass transition temperature T_g. For dry PET it is about 72°C. If the temperature continues to rise, at around 100°C the PET becomes tacky, but doesn't melt entirely until the temperature exceeds 250°C. In contrast, when crystalline PET is heated, there are no rubbery or tacky phases, and in fact the material barely softens before melting at about 250°C (which has important implications when drying PET).

As they cool, most polymers shrink slightly. For amorphous thermoplastics, the shrinkage is mainly due to the part contracting inside the mould as it cools, but with a semi-crystalline thermoplastic such as PET, shrinkage also takes place as the polymer crystallizes and the molecular chains become more densely packed. The degree of crystallinity is a function of the cooling rate, and so thick parts which retain their heat for longer tend to be more highly crystalline and thus shrink more. Certain pigments can also affect the crystallisation rate (and

---

[3] Crystallinity is typically quoted in terms of the proportion of the PET in a sample that has crystallized, and is determined indirectly from its density as measured in a Density Column. A Density Column is made by carefully mixing in a clear cylindrical glass tube, two miscible liquids having a density range in excess of the range of density to be investigated, such that the density of the resulting mixture varies linearly along its length, with the lowest density being at the top and the highest at the bottom of the tube. Small calibrated glass floats of precisely known density are introduced into the top of the column and sink to the level where their density matches that of the solution. To use it, tiny PET samples are cut from the container wall and introduced into the top of the column, where they sink until they reach liquid of their own density and therefore stop sinking. The density of the sample is then determined by reference to the nearest calibrated glass float.

Percentage crystallinity = $(\rho_s - \rho_a)/(\rho_c - \rho_a) \times 100$

Where ρ_s is the measured density of the sample, ρ_a = 1.333, the density of amorphous PET, and ρ_c = 1.455, the theoretical density of PET crystal calculated from unit cell parameters.

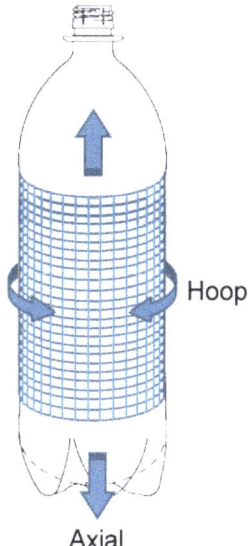

consequently the shrinkage) since they nucleate ("seed") crystal growth, thus making it essential to achieve a homogeneous dispersion of the pigment in the polymer matrix.

No polymer can be 100% crystalline because the chains (on a microscopic level) are too "long" and tangled for perfect alignment. PET can have crystallinity levels between 0 and 55%.

Orientation

If sufficient stress is applied to a ductile material it will stretch. Initially the stretching will be reversible or *elastic*, which means that if the stress is removed the specimen will return to its original length. But sooner or later a yield point will be reached where permanent deformation remains even after the stress is removed, i.e. the material has been stretched beyond its elastic limit. If the stress continues to increase, the material will carry on stretching until it ultimately breaks.

- Stress — force per unit area applied to a body that causes it to deform. It is a measure of the internal forces on the molecules as they resist separation, compression, or sliding in response to externally applied forces. Tensile stress is the axial force per unit area applied to a body that tends to extend it linearly.

- Strain — measure of the extent to which a body is deformed when it is subjected to a stress. The tensile strain is the ratio of the change in length to the original length.

- Elastic Modulus — ratio of the stress applied to a body to the resultant strain.

PET has another property: in common with a limited number of polymers including PA (polyamide, better known as Nylon), and PP (polypropylene), if warmed to a temperature where its chain-like molecules are sufficiently mobile to *unfold* instead of breaking as the material is

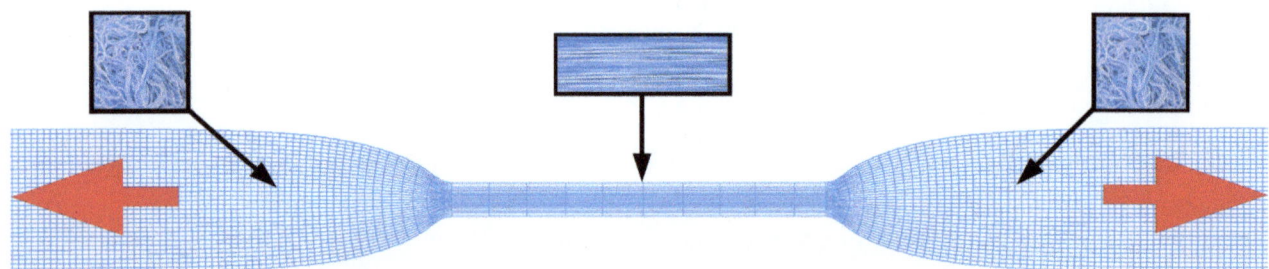

Localized necking occurs once the yield point has been attained
(smaller images represent the molecular structure)

stretched, when the yield point is exceeded PET can be strain hardened or *oriented*. Oriented PET consists of closely packed molecular chains aligned in the directions of stretch and is stronger than amorphous PET because the molecules work together to support loads that would break individual chains. *Biaxial* orientation results when stretching is applied from two directions at right angles like when a preform is stretch-blown to form a bottle.

Of course such an extended molecular conformation wouldn't normally be stable, but the instant the expanding preform touches the cooler blow mould surface, the chains are frozen in their extended/ oriented arrangement and a rigid container results. (Except if the container should happen to be heated to a temperature at which the PET chains are once again able to move, when it will shrink—that is if the "memory" of their original arrangement has not already been removed by annealing).

Stretching also causes segments of adjacent chains to line up in a close-packed regular conformation akin to a crystalline array. Known as *crystallites*, their small size means they don't impair transparency, but their sheer numbers significantly improve the physical properties of the material because they help to anchor the extended macromolecules in place.

PET can exist in at least three physical states:

1   Amorphous — e.g. the preform, and the neck and base area immediately adjacent to the gate in the bottle, . The preform is amorphous because there is not sufficient time allowed in the injection mould for the molecules to crystallize.

2   Amorphous-crystallized (amorphous sections having thermally induced crystallization) — e.g. in the gate itself

3   Oriented (mechanically induced crystallization: more orientation yields more crystallization) — e.g. the bottle sidewall

Orientation, where it exists, exerts a powerful self healing effect leading to a uniform wall thickness in a bottle. Consider the stretch-blow moulding process: if we suppose the stretching force to be transmitted uniformly throughout the preform, by definition the tensile stress will be highest at the narrowest cross-section. In consequence this region stretches first and forms a "neck", but it will only stretch by as far as is necessary to attain the natural stretch ratio and thus make the material in this region stronger than the material in the adjacent region— which will then stretch in turn. This process repeats with stretching progressing along the preform; and if the stretch ratios are high enough to achieve the natural stretch ratio in every part of the preform, gives a bottle with a uniform wall thickness. Without strain hardening, such a self-healing process would not occur and any local non-uniformity might lead to catastrophic cross-section reduction and

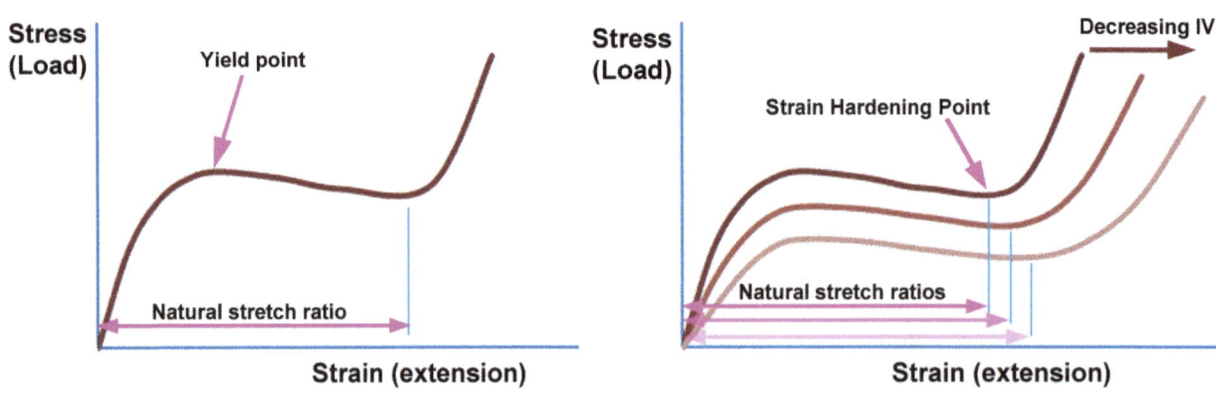

Polyethylene Terephthalate (PET)

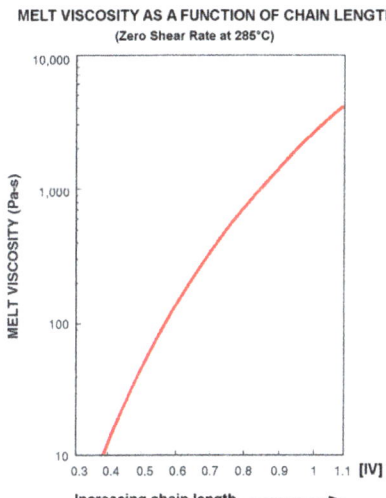

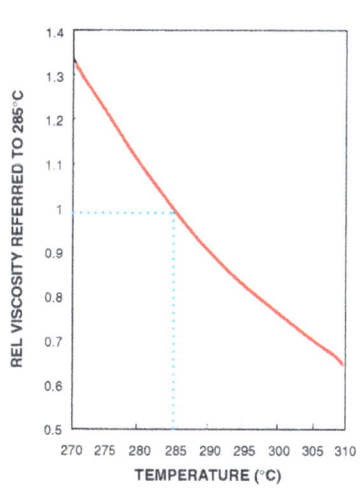

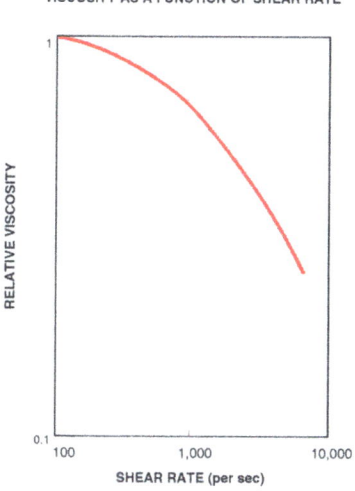

break. Of course high stretch ratios are ultimately required to achieve the uniformity, and so ideally the preform dimensions and wall thickness will be carefully attuned to the container shape.

The factors controlling the degree of orientation are (a) temperature, (b) strain rate, (c) crystallization, (d) moisture content, and (e) molecular weight. If PET is stretched while it is too cold, the force required is very high and the material ruptures (perceptable as *pearlescence*). At higher temperatures the polymer will flow more and so a much greater stretch is required to reach the strain-hardening point (the *natural stretch ratio*). Conversely at lower temperatures the natural stretch ratio is smaller. At high strain rates (stretch speeds) there is considerable molecular resistance to chain disentanglement and movement and so strain hardening occurs at lower stretch ratios (although stretching forces are higher). It is essential for the preform to be amorphous; crystalline PET cannot be stretched without breaking.

Water acts as a plasticizer in PET. Preforms stored in humid conditions will absorb moisture, the effect of which is to lower the glass transition temperature. At high relative humidities (RH > 90%) the T_g can be reduced by several degrees, which has a pronounced effect on the blowing process and the product properties. The effect is similar to blowing at higher temperatures as the polymer will flow more easily. Of course this is not a problem in injection stretch-blow moulding where the preforms are blown immediately after they have been moulded and thus don't have time to absorb moisture from the atmosphere, but in a two-stage operation it is important to equilibrate the preforms at constant RH prior to blowing.

Intrinsic Viscosity

The chain-like molecular structure of PET not only provides the characteristic plasticity, but also influences the flow properties of the melt: the longer the chains (i.e. the higher the molecular weight), the greater the viscosity (although things are complicated by the fact that at the high shear rates experienced when melt is injected through the barrel nozzle, the molecules have a tendency to disentangle, resulting in an apparent reduction in the viscosity). It being impracticable at the time to directly count the number of repeat units, instead of categorising PET grades by their chain length, the *intrinsic* viscosity (IV)[4] was used instead. The term intrinsic viscosity refers to the ratio of the specific viscosity of a polymer solution of known concentration to the concentration of solute, extrapolated to zero concentration. Its value is directly proportional to average polymer molecular weight.

Application	Repeat units	Molecular Weight (average)	Intrinsic Viscosity
Fibre	105	20,000	0.6
Film	120	22,000	0.65
Generic bottles	130	25,000	0.75
CSD bottles	146	28,000	0.84
Tyre Cord	167	32,000	1.0

---

[4] Intrinsic Viscosity used to be determined by plotting $(\eta - \eta_o)/\eta_o.C$ against concentration C in g/ decilitre and noting its extrapolated value to zero dilution hence *intrinsic* viscosity; where η is the viscosity of the polymer in solution, and η_o is the viscosity of the pure solvent (traditionally a 37.5%/ 62.5% by weight solution of Tetrachloroethane in Phenol). Nowadays polymer molecular weights can be determined directly by Gel Permeation Chromatography but the resulting number is still converted back into the familiar IV.

In fact, unlike with smaller molecules, the molecular weight of a polymer is not one unique value. All synthetic polymers are *polydisperse* meaning they contain polymer chains of different lengths, and so the molecular weight is not a single value — the polymer exists as a distribution of chain lengths/ molecular weights depending on the way the polymer was produced. The importance of the IV is that it is a unique value which embodies the molecular weight *distribution* of each resin.

How PET is made

In order to produce PET resin, a *terephthalate* component and a *diol* component are reacted in the presence of a catalyst to form the *monomer* (with water as a by-product). The term "diol component" refers primarily to mono-ethylene glycol (MEG — anti-freeze), although other diols can be added to form a *co-polymer* (see below). A variety of catalysts can be used, including antimony, germanium, or cobalt. Although a catalyst is a substance that increases the rate of a chemical reaction without itself undergoing any change, in practice minuscule traces remain in the PET to give each material its characteristic tint.

The monomer is then heated in the absence of oxygen to initiate *condensation polymerisation* yielding a macromolecule with around 100 repeat units. Ethylene glycol is a by-product of this process, and must be continuously removed to maintain favourable reaction conditions. In order to cool the strands very rapidly, the molten polymer is then extruded through a multi-orifice die into a water stream where the filaments are chopped by rotating blades into pellets the size of a grain of rice, which are then dried in a cyclone separator.

The polyester thus formed typically has an intrinsic viscosity of between 0.60 dl/g and 0.65 dl/g, which is adequate for fibre applications, but its low melt strength and fast crystallization make it difficult to process on conventional stretch-blow moulding machines. Furthermore the pellets are amorphous (meaning they would become tacky and clog the material hopper at normal drying temperatures), and they contain acetaldehyde (AA). In order to produce high-clarity containers, PET needs to have an intrinsic viscosity of between 0.72 dl/g and 0.84 dl/g, and so in a final step the polyester is further polymerized in a process known as *Solid State Polymerisation*. In order to make the medium length chains grow longer, the pellets are warmed in an inert atmosphere; the longer this warming continues, the higher the IV. Beneficial side effects of this warming include the amorphous pellets crystallizing (crystalline PET is not tacky at drying temperatures), and a portion of the retained AA evaporating.

It is not only the melt strength which increases as the molecular chains get longer (IV increases), the molten polymer is more viscous, meaning higher processing temperatures are needed, and to stretch the preform requires a larger force. The increased stretch force requirement of high IV material enhances the self-healing tendency of PET undergoing biaxial orientation leading to a more uniform material distribution in the container (although any improvement may be limited by the need for the preform to be hotter when blown in order to avoid *pearlescence*).

Not only is higher IV resin more expensive, but its processing entails higher energy usage, a longer cycle time, and a greater potential for polymer degradation. However its advantages outweigh these drawbacks; by using a higher IV resin the toughness, impact strength, and barrier properties of the bottle are improved, as well as its environmental stress cracking (ESC) resistance. All of which explains why high IV resins are recommended for making carbonated soft drink containers.

Most commercial PET polymers are actually *co-polymers* in which low levels of modifiers have been randomly substituted in the polymer chain as a "co-monomer". This substitution disrupts crystallization, thereby improving clarity and altering the melting point. Commonly used modifiers include isophthalic acid (IPA), diethylene glycol (DEG), and cyclo-hexane dimethanol (CHDM), although the latter reduces oxygen barrier.

PET bottle resin is a commodity product, priced according to demand and the availability of its precursors: paraxylene (the feedstock for the manufacture of PTA) and MEG (mono ethylene glycol), both of which are ultimately derived from oil and priced accordingly, as can clearly be seen from the chart below.

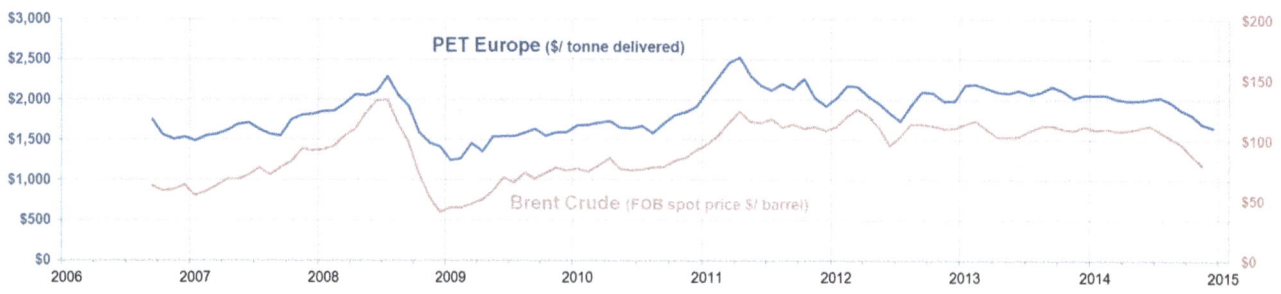

Container design

decide upon the container's capacity, neck finish and overall shape, and calculate the surface area of the body (excluding the neck)

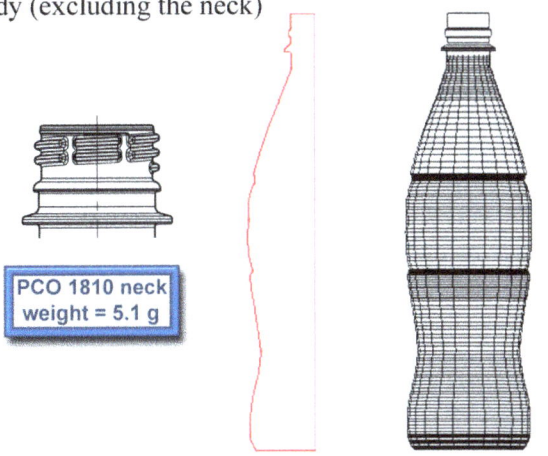

PCO 1810 neck weight = 5.1 g

Decide upon the weight

first determine the volume of PET in the container body by multiplying the calculated surface area by the average wall thickness that past experience suggests will give a container with the required performance

then multiply the calculated volume by the average density (1.37g/cm³) of the PET found in a typical container body (a combination of oriented and amorphous material) to find the preform body weight (add the preform body weight to the weight of the neck to find the total preform weight)

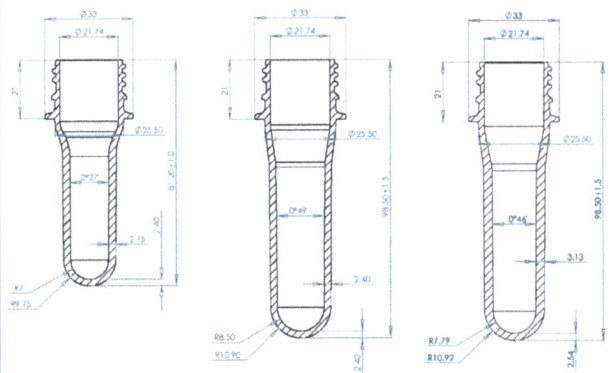

Preform design

calculate the volume of the preform body by multiplying its weight by the specific volume of amorphous PET: 0.75 cm³/g

starting with the neck finish, sketch a preform body having the optimum stretch ratios. Calculate the volume of the body, and if necessary adjust its dimensions (by compromising on the stretch ratios) in order to give the previously determined volume

the maximum preform wall thickness should not exceed 4.5 mm, and ideally will be less than 4 mm

Reheat additives

Another range of additives — common in bottle-grade PET but actually irrelevant to injection stretch-blow moulding — enhances *reheat capability*. In the two-stage process the injection moulded preforms are *reheat blow moulded*. The preform is heated to the correct temperature by infrared halogen heat lamps before it is stretch-blown inside a blow mould. The time required for the preform to reach the right temperature is known as the reheat time and is a function of the absorption characteristics of the resin. The heat up time limits the number of preforms that can be blown per hour on a machine, making it desirable to provide resins that reheat faster and/or require less energy. A variety of different additives can be used to enhance the heat absorbing properties of PET including finely divided carbon black, metal particles, and even coloured dies. All are black- or grey-body absorbers that absorb energy across the entire structure of infrared and visible radiation — which means they impart a tint to the polymer in proportion to the amount added.

Drying

Many polymers are hygroscopic, meaning they absorb water from the atmosphere. At ambient temperatures this usually doesn't do them any harm, but if moisture happens to be present when the polymer is melted, it can lead to problems. First, the entrained moisture may be released from the melt inside the mould cavity, resulting in surface or interior defects. Second, and potentially most damaging, when PET is heated a chemical reaction with the moisture (*hydrolysis*) results in the breakup of the molecular chains, ultimately reducing the physical properties of the material. The scale of the IV reduction is a function of the quantity of moisture that is present, which means that in order to produce consistent containers, reducing it to an acceptable level by pre-drying is essential. (Read more in *Chapter 4—Drying*).

Non-hygroscopic polymers such as polypropylene do not absorb moisture because the space between their molecules is smaller than the size of a water molecule, and therefore typically don't need drying.

Use of regrind

An important factor affecting the preform IV is the proportion of *regrind* blended with the virgin PET pellets. Since the regrind IV is generally lower than that of virgin PET, the amount used must be strictly maintained at a constant level (≤ 20%), otherwise the fluctuating IV of the preforms could lead to variation in the container wall thickness and in severe cases, the production of bottles with a crystalline haze. If there is any question about the proportions or consistency of the blend, it is usually best to empty the dryer hopper and refill it with 100% virgin pellets. It is also important that the flake is clean and pure, and is uniformly mixed with the virgin pellets, with no "clumping" of the regrind.

Acetaldehyde

A substance produced in small amounts during the melt processing of PET. At room temperature it is a colourless vapour that is very soluble in water and has a sharp, penetrating fruity smell. It is widely used in the food industry as a flavour ingredient. However acetaldehyde generated during the melt-phase polymerisation of PET can be trapped in the amorphous material during the cooling and pelletising operation. Some of this trapped AA is driven off in the solid-phase polymerisation process, but this still leaves a residual level of around 1.5 ppm in the virgin pellets. More will be driven off when the PET is dried prior to moulding, but the injection moulding process usually generates additional AA which is then trapped in the preform and ultimately the bottle wall. The capacity of AA to diffuse out of a bottle wall and impart a flavour to the contents means that its generation during preform moulding must be carefully controlled (mineral waters and cola-type beverages are particularly sensitive to tainting). The generation of AA is not associated with any significant IV loss but is a thermal decomposition reaction. To minimise the amount generated adopt the mildest possible moulding conditions to deliver the lowest practicable melt temperature:

- Reduce the barrel set temperature (although if the temperature indicators show the barrel heaters are being overridden because of shear heat from the screw, reducing the settings will have little effect)
- Reduce the back pressure on the screw (a little may be necessary to achieve a homogeneous dispersion if pigments are being added)
- Reduce the screw speed (if possible until the "dead time" between the completion of injection barrel charging and the start of the next injection stroke, is at a minimum)

Other factors that will help to keep AA levels low include injecting the molten PET into the cavities at a slower rate to reduce the shear heating, using the minimum cushion of material forward of the screw, and keeping the temperature of the hot runner system as low as possible.

Preform design principles

A preform's design is dictated by the container parameters (size, shape, neck diameter, specific requirements of the application) and the natural stretch ratio of the PET (as influenced by the stretch-blow moulding process condtions).

Ideally a preform should be designed to have a stretch ratio that is somewhat higher than the natural stretch ratio of the polymer. By stretching it beyond its strain hardening point, the orientation and crystallization can be optimized and the performance of the resin maximized.

The *axial* stretch ratio is defined as the length of the stretched part of the preform divided by the height of the same part of a bottle. The *hoop* stretch ratio is defined as the bottle diameter divided by the preform diameter, but since preforms are tapered, judgment is required when deciding at which part of the preform the diameter should be measured. Similar problems arise with bottles having large

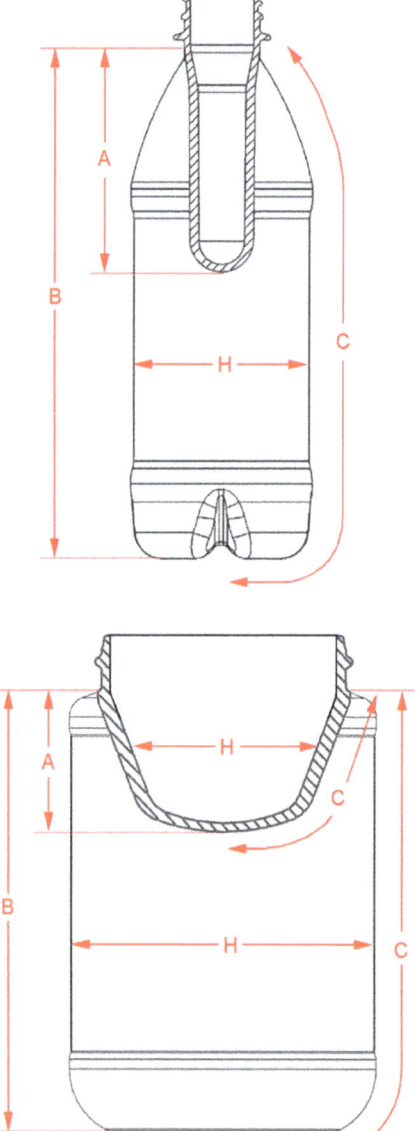

A = preform axial stretch length
B = container axial stretch length
C = contour axial stretch length
H = hoop diameter

changes in section, and also with oval containers.

In such cases it can be better to determine the hoop stretch ratio by comparing the preform and bottle circumferences. The overall or *area stretch ratio* is the product of the hoop stretch ratio and the axial stretch ratio.

The natural stretch ratios of various polymers have been determined by experiment:

Polymer	Hoop	Axial	Area
PET	3.0-4.0	2.0-2.5	10
PET (CSD)	5.6-6.5	2.4-2.7	15
PP	2.5-3.5	2.0-4.0	10
PC	2.0	1.5	3

Note that if the axial stretch ratio is low, then the hoop ratio should generally be higher, and *vice versa*. In broad terms using lower stretch ratios means that the material can be more easily blown into the heel of the container, but the bottle wall thickness will likely be more variable. PET bottles designed for carbonated soft drinks (CSD) require higher levels of orientation and so their preforms should be designed with higher levels of stretching. It is notable that PP stretch ratios typically result in short, thick-wall preforms, while PC stretch ratios result in longer, thinner-wall preforms.

Preform dimensions can seldom be determined solely in order to maximise the stretching since other factors such as the container and preform wall thicknesses also have to be considered. For example automatic filling and capping machines often apply a significant downward force on a container; also filled PET containers are normally stacked with several layers on top of each other, thus imposing a significant load on the ones at the bottom. What this means is that if it doesn't have sufficient top load strength, a container will collapse, either when being filled, capped, stored, or transported.

Consumers are also influenced by the feel of the container when gripped. If sidewall deflection is too high, it can be perceived as a poor quality product (to avoid this problem it may be helpful to assess the wall thickness of comparable containers). Furthermore PET containers that are used to contain and store carbonated beverages in warm climates and/or during the summer months undergo thermal expansion, or *creep*, caused by the pressure generated in the container at high ambient temperatures. Creep increases the volume of the container, thus lowering the fill height and making it more likely that the consumer will reject the container as underfilled.

Top load strength, creep resistance, and sidewall regidity all depend upon wall thickness (as well as the degree and uniformity of the biaxial orientation). Next it is necessary to take into account the thickness of the preform wall implicit in the chosen average container wall thickness. With conventional bottle-grade resins the maximum preform wall thickness should not exceed 4.5 mm (above this the low thermal conductivity of polymers will prevent effective cooling, resulting in crystalline haze in PET preforms). Specially formulated slow-crystallizing PET grades have been developed for thick-wall applications, but the thicker the wall, the more expensive the container will be to make. Partly due to the cost of the resin, but also owing to the cost of the energy consumed during production, which is directly related to the thickness of the preform (because in a thicker preform there is more polymer present to heat and cool). Very often the ideal stretch ratios cannot be used because an overriding requirement for a faster cycle demands a thinner wall, which for a given weight necessitates a wider or longer preform.

Heat Setting

Certain products such as fruit juice and dairy products need to be pasteurised to prolong their shelf life. The bottle containing the juice also needs to be sterilised and so methods have been developed of flash pasteurising the juice and then bottling it while still hot in order to sterilise the container at the same time. Standard PET bottles are unsuitable for hot filling applications because when heated above approximately 60°C they soften and, as the chains attempt to re-fold, shrink. At best this will mean fill levels are incorrect, but in more extreme cases the bottle becomes misshapen. Why does this happen? As a preform is stretch-blow moulded, the chain-like PET molecules unfold, and then before they can revert to their original configuration, contact the blow mould and are frozen into their new distended arrangement. However this molecular arrangement is unstable, and should the PET temperature approach the T_g, the polymer chains have a tendency to relax and revert to their original configuration, i.e. the bottle gets smaller. This "memory" may be minimised by (i) blowing the preform while very hot (although this means the degree of orientation is reduced), or (ii) increasing the crystallinity of the PET (since crystalline material doesn't relax and also the crystals help to "anchor" the oriented molecular chains in place). The generic term for the application of heat energy to raise the level of crystallinity in a PET bottle in order to improve its thermal stability is "heat setting".

PET crystals are either formed mechanically through orientation, or thermally through the application of heat energy. *Mechanical crystallization* of PET is a by-product of orientation and results in many very small "crystallites", whereas *thermal crystallization* of PET depends upon both time and temperature and tends to result in fewer, larger crystals. Several other factors can also influence the degree of crystallinity, such as resin grade (copolymers with slow crystallization rates have been developed), intrinsic viscosity (high IV grades crystallise more slowly), or the presence of nucleating agents designed to promote the growth of many small crystals), and so an appropriate PET resin grade must be selected.

Previously it was stated that PET can exist in three different physical states: amorphous, oriented, and amorphous-crystallized. It can now be seen that when it is "heat set", PET exists in a fourth state: *oriented-crystallized* (oriented chains that have been further crystallized by thermal energy).

The earliest heat setting techniques were "single blow" methods (as opposed to later "double blow" techniques developed by Sidel and Nissei ASB), in which bottles are blown and then held for a short time inside heated blow moulds to both enable the tiny strain-induced crystallites to grow, and to relax some of the stresses induced in the PET by the stretching process (stresses that might otherwise cause the bottles to shrink when hot filled).

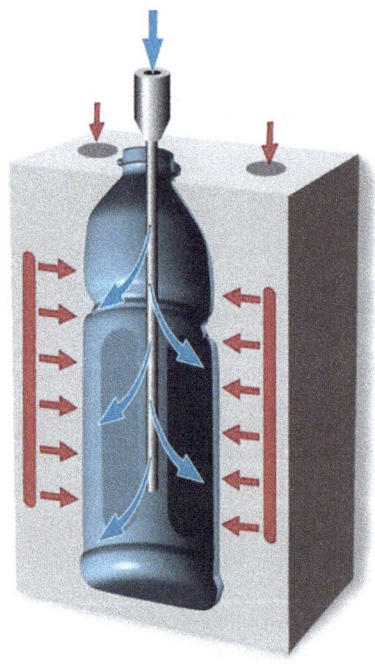

The degree of improvement that can be achieved by heat setting depends on the blow mould temperature. The maximum temperature is limited by the need for bottles leaving the mould to be rigid and thus resistant to deformation, and so simple single blow methods (blow mould temperature ≤100°C) can only raise PET bottle service temperatures to about 80°C. Enhanced single blow methods (blow mould temperature ≤125°C) incorporating in-mould cooling can raise service temperatures after storage to around 83°C with standard grades of PET. In-mould cooling involves blowing cold air through a hollow stretch rod, closed at the bottom end and cross-drilled to create a pattern of cooling on the inside surface of the bottle). It is sometimes referred to by the French word *balayage* which means "to sweep" [i.e. sweeping a jet of cooling air around the inside of the container].

The objective when heat setting PET is to maximise the level of crystallinity in the bottle wall without reducing clarity. There are certain fundamental principles to be followed:

a) Preform design, temperature, and stretch conditions should be selected to induce the maximum orientation.

b) The preform must be blown into and then held under pressure in a hot blow mould for crystallites to form, which means the blow cycle time will be longer than normal..

During bottle storage the moisture absorbed by the PET will decrease its hot fill capability and also reduce the glass transition temperature (Tg). The Tg is important because at temperatures above the Tg, PET starts to lose its stiffness. This means that moisture absorption cannot simply be neglected but must be taken into account when comparing heat set PET bottles (and quoting a "safe working temperature" for a heat set PET bottle means little if the storage conditions are not also specified).

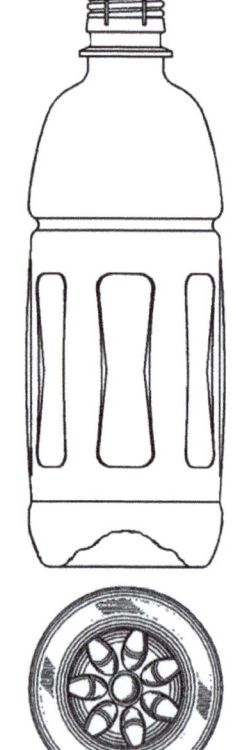

The shape of the upper part of this PET bottle intended for hot fill applications is constructed using curved surfaces to make it very rigid in order to prevent it from crumpling as a result of the partial vacuum created inside when the contents cool and contract. The lower half of the bottle has movable panels in the sidewall to "accommodate" this partial vacuum without visibly creasing.

The concave base of the bottle is shaped so as to make its surface area as large as possibe. This promotes stretching to minimize the amount of amorphous, and maximise the amount of oriented material in this area.

Immediately after the heat set bottle has been filled and its closure applied, the container is inverted in order to sterilize the inside of the neck, but this introduces another problem: only the body of the bottle is heat set. The neck remains amorphous and so liable to distort at high temperatures. There are two ways to overcome this. Either the necks are made very thick and the filled containers cooled as quickly as possible, or the preform neck may be crystallized after blowing. This involves heating the neck to a temperature high enough to induce crystallization, typically 160°C. PET bottle neck crystallisation systems are available from Osaka Reiken Co. Ltd of Japan.

Heat setting adds substantially to the container cost both in terms of reduced output per mould and the increased capital cost.

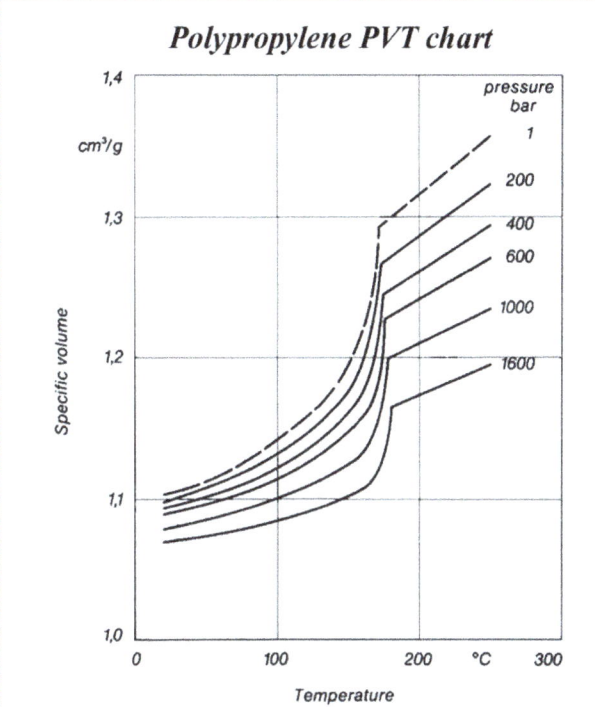

The relationship between pressure, volume, and temperature is shown graphically in this PVT chart for polypropylene. Measured during heating up, the number of cubic centimetres required to accommodate each gram of polypropylene increases sharply as the crystals melt. (Note that specific volume is the *volume per unit of mass* and is therefore the reciprocal of density).

PP

PP copolymer has reasonable impact strength, although it is brittle below 10°C. It is fairly clear with a good surface gloss and is not harmed by boiling water. Its lower stiffness in comparison to PET means that PP bottles made with similar wall thickness will have lower top-load and hoop strength, and so container walls may need to be made thicker in order to achieve the required stiffness. Since PP has a lower specific density than PET (0.9 as opposed to 1.34), despite needing thicker walls, comparable containers made from the two materials can often have similar weights.

Thermoplastics expand when they get hot and shrink as they cool down, but for semi-crystalline plastics such as PET and PP the situation is more complicated because crystalline regions shrink more than do amorphous regions. Crystalline materials go through a much larger volumetric change because the organized crystal structure takes up much less space than the disorganized amorphous regions. The relationship between pressure, volume and temperature can be shown graphically in the form of a PVT plot which shows a discontinuity in the curve which corresponds to the region of crystalline fusion.

PP can be oriented, but needs to be stretched much further (~25%) than does PET which may not be feasible. Without the automatic wall thickness control imparted to PET by biaxial orientation, attaining an even wall thickness in a PP container therefore depends utterly on the accuracy of the preform mould and the uniformity of the preform heat content. As mentioned above, PP bottles—and thus of course their preforms—must be made thicker, but because thick wall sections retain the heat for longer, they contract significantly as they cool, leaving those areas of the preform where the hot plastic has shrunk away from the cavity hotter than those parts where shrinkage has been less. In partial compensation for this, it can be advantageous to set a longer than average waiting time prior to blowing to

	PP	PC	PET	PES	Tritan™
Specific density [g/cm³]	0.9	1.2	1.34	1.36	1.18
Melt density* [g/cm³]	0.85	1.01	1.15	1.34	
Tensile strength @ Yield [MPa]	25	62	54	70	43
Flexural modulus [MPa]	1700	2400	2600	2700	1575
Water Vapour Transmission**	9-11		16-20		
Deflection temperature at 1.8 MPa [°C]	53	130	74	174	81
Drying temperature [°C]/ time	N/A	120 / 4 hr	150 / 4 hr	150 / 4 hr	88 / 4 hr
Barrel temperatures [°C]	215-260	280-300	270-285	355-365	260-282
Injection Mould temp [°C]	15-65	80-90	10-15	140-160	38-60
Blow Mould temp [°C]	35-60	35-60	10-25	150	

*Melt density at typical processing temperatures.

*Normalized WVTR @ 38°C, 90% RH for 1 mil (0.001") film. Units: g/m²/24 hr. The WVTR for a specific thickness may be calculated by dividing the value found in the table by the actual gauge in mils (thousandths of an inch).

allow time for the heat profile to even out. The conditioning station on a four station injection stretch-blow moulding machine can be helpful here, but for efficient heat transfer the preform needs to actually touch inside the heating pot (which must therefore be polished to a mirror-like surface finish). To avoid creating scratch marks on the outside of the preform when it is withdrawn, once the heating pot is in position low-pressure air is used to very slightly expand the preform until it gently presses against the inside of the pot, (the amount of expansion is determined so that the preform shrinks back elastically to its original dimensions when the air is released). To allow for the possibility of cooling as well as heating the preform, the pot temperature should be adjusted by the circulation of thermo-oil supplied from an external Temperature Controller.

With OPP containers clarity depends upon:

- the degree of stretching (i.e. orientation). PP needs to be stretched further than PET. Parts of the preform such as the base and neck that experience less orientation will always be hazy.
- the level of crystallinity (which can be improved through the addition of nucleating agents)
- the wall thickness (not surprisingly the thicker the wall, the worse is the clarity)
- although transparency may appear poor upon first sight, the contact clarity can still be remarkably good (the apparent improvement is greater for coloured products)

Polypropylene is a semi-crystalline polymer. The crystals enhance the stiffness, as well as the mechanical, chemical and thermal resistance of the material. PP normally crystallizes slowly and forms relatively large complex crystal aggregates known as spherulites. These are generally larger than the wavelength of visible light, and so scatter it, causing haze. By adding nucleating agents (clarifiers) to PP, many additional sites are created for spherulites to grow. Because many more crystals are growing in the same amount of space, they are all much smaller in size. The result is large numbers of crystals smaller than the wavelength of visible light that allow light to pass through, thereby improving clarity. Most PP grades intended for stretch-blow moulding are supplied ready-compounded with a nucleating agent.

PES

Classic one-stage machines are designed to process PET which has a relatively low melt temperature in comparison to the 340–390°C of polyethersulfone (PES). The machines are designed to operate with colder injection and blow moulds than the 112–125°C and ~135°C respectively needed for PES. Before moulding this high temperature polymer it is necessary to ensure that the barrel is fitted with ceramic heater bands capable of maintaining the required temperatures, that the preform and blow moulds are thermally insulated from the machine bed, and that they are connected to Mould Temperature Control units capable of providing an adequate fluid flow at the required temperatures (as high as 160°C for the preform, while the blow mould will need to be almost as hot).

- The resin dryer:
  PES material must be dried prior to moulding in order to reduce the water content to 0.05% or below to prevent deterioration of the resin/ the formation of surface defects. PES can be dried using a dehumidifying hopper dryer for 3 to 5 hours at a temperature of 150 to 170°C.
- The one-stage machine:
  Normal advice is to choose the injection barrel so that the shot weight lies between 30 and 70% of the injection capacity, but to minimise the residence time when moulding PES, if possible the barrel should be selected so that the shot weight is closer to its theoretical maximum.
- The hot runner system:
  A flow-balanced cavity arrangement in which all cavities fill at the same rate with the same pressure (achieved by placing the cavities equidistant from the inlet and having identically sized runners) is crucial for moulding uniform preforms. Hot runner designs with long residence times or dead spots where material can stagnate and degrade must be avoided at all costs. Manifold channels should be free-flowing without sharp corners or flow impediments because flow restrictions will increase the shear on the material which may cause discoloration and degradation of the resin. Open nozzles are preferred to nozzles equipped with shut-off devices. Ceramic band heaters are recommended because they are better suited for high temperature use, provide a higher watt density than mica band heaters, and typically last longer. Gates must be of adequate size to allow part filling without the use of excessively high melt temperatures or pressures.
- Stoppages/ Shut down:
  At normal processing temperatures the sulfone resins can tolerate a residence time of no more than 10 to 20 minutes, which means that if processing were to be suspended for longer than this, the temperature must be reduced without delay, while for a longer shut down, it may be best to purge the PES completely from the barrel.

4

Drying

PET resin is hygroscopic, meaning it absorbs water from the atmosphere. At ambient temperatures this doesn't do any harm, but if too much moisture is present during processing, it will lead to problems. Molten PET reacts with water in a chemical reaction called *hydrolysis* which ends up with the macromolecule breaking in two, in other words the IV is reduced. The lower the preform IV, the more difficult it is to blow a bottle with an acceptable material distribution and the faster crystalline haze will appear (once the IV falls below 0.65, it's impossible to make clear preforms). The magnitude of the reduction is a function of the moisture level. Ideally this should be less than 40 ppm (parts per million) or 0.004% by weight, so that as shown in the graph below, any IV reduction cannot be greater than 0.04. The graph also highlights that the lower its IV, the less susceptible is PET to hydrolysis.

Hydrolysis reduces the IV of PET. The magnitude of the reduction being related to the level of moisture in the resin. The lower the IV, the less susceptible is the resin to the damaging effects of hydrolysis.

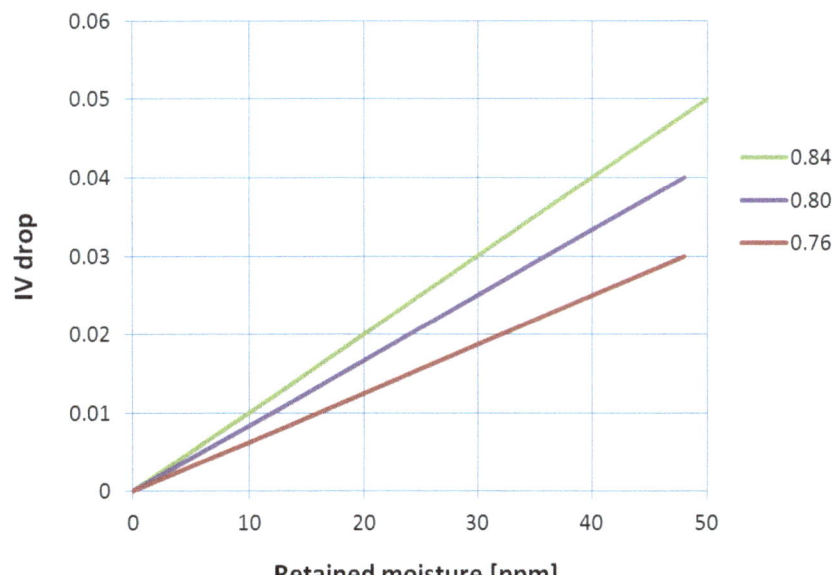

Dew Point		Vapour Pressure (Water/Ice in Equilibrium)		PPM on volume basis at a pressure of 760mm Hg/ 1 Bar	Relative Humidity at 70°F/ 21°C %	PPM on weight basis in air
°C	°F	mm of Hg (Mercury)	Bar			
-40	-40	0.09660	0.00013	127	0.516	78.9
-30	-22	0.28590	0.00038	376	1.52	234
-20	-4	0.77600	0.00103	1020	4.13	633
-10	14	1.95000	0.00260	2570	10.4	1596
0	32	4.57900	0.00610	6020	24.4	3640

Note that since "parts-per-million" fractions are quantity-per-quantity measures, the units of measurement cancel each other out, making them pure numbers with no associated units of measurement. Incidentally, one part per hundred is of course usually represented by the percent (%) symbol and so 1% denotes one part per hundred parts, or ten thousand parts per million parts (being the same fraction), which means for example that 0.004% is exactly the same as 40 ppm.

The outcome on the stretch-blow moulding process of having the incorrect IV is illustrated in this diagram:

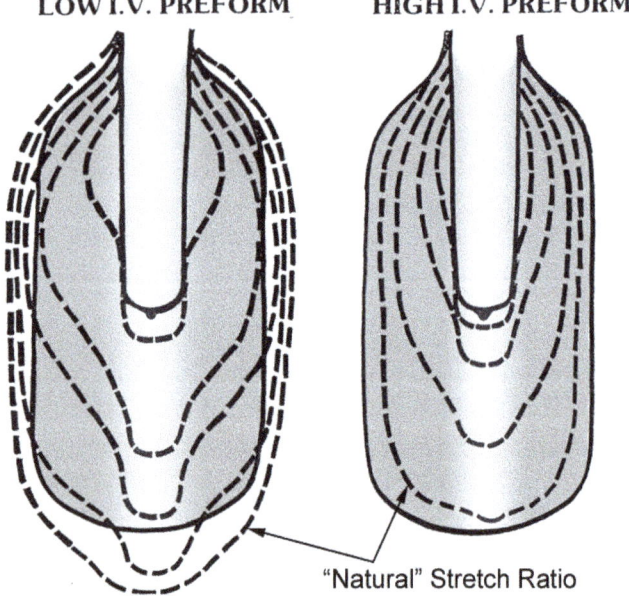

"Natural" Stretch Ratio

The righthand image shows the optimal high IV preform being stretched inside a (shaded) blow mould and represents the ideal situation, but in the image on the left, the low IV preform stretches too easily and since the material clearly cannot be stretched beyond the walls of the blow mould, the most likely outcome is a deep fold (or *crater*) in the base of the bottle.

Drying

Proper drying is the most critical aspect of moulding good quality PET preforms. Water absorbed by PET diffuses throughout each pellet until an equilibrium with the surroundings is achieved. The aim of the drying process is to reverse the diffusion and make the moisture leave the pellet. An understanding of what influences drying efficiency will be helpful. Consider a liquid held in a closed container: it will evaporate until an equilibrium is attained where as many molecules return to the liquid as escape. At this point the vapour above the liquid is said to be saturated, and its pressure is the saturated vapour pressure. If the liquid is open to the air, then the vapour pressure exerts a partial pressure along with the other constituents of the air. (A partial pressure is the pressure one gas in a mixture of gases would exert if it were the only one present).

At higher temperatures, more molecules have enough energy to escape from the liquid and the saturated vapour pressure increases. The temperature of the liquid at which its vapour pressure is equal to the atmospheric pressure (thus enabling bubbles of water vapour to form inside the liquid), is its boiling point. (Incidentally, what is often referred to as steam is actually free floating droplets of liquid condensed out of the invisible water vapour).

Boiling and evaporation are two different processes: with evaporation liquid gradually vaporises at temperatures below the boiling point. The rate of evaporation depends upon:

- Air flow: the higher the air flow, the more evaporation
- Temperature: the higher the temperature, the more evaporation
- Evaporation is a surface phenomenon: the larger the surface area, the more evaporation.
- Humidity: the lower the humidity, the more evaporation

The relative humidity relates directly to the partial pressure of the water vapour within humid air, being the ratio of the amount of water vapour that is actually in the air, to the amount of water vapour that would be present in saturated air at the same temperature. For effective drying it is necessary to have air that is not saturated, since the lower the relative humidity, the greater is the capacity of the air to hold moisture.

All of which means that to properly dry PET pellets, they must be held at an elevated temperature for a sufficient length of time, while being bathed by

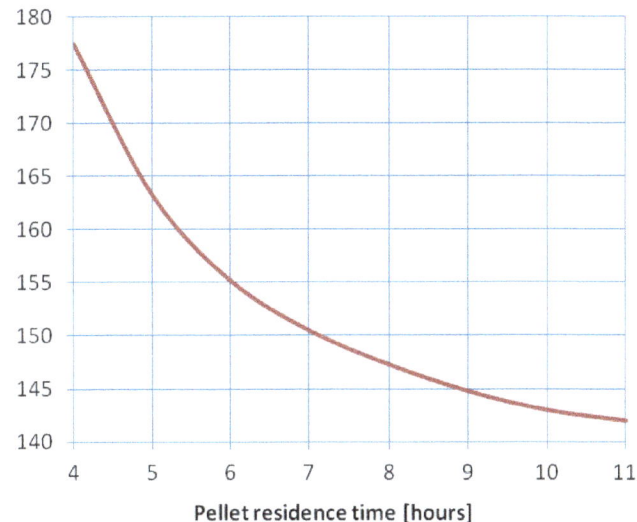

Drying temperature versus time

a flow of dry air sufficient to carry away the moisture that is evaporating from their surface.

In a typical PET dryer, air is first dehumidified to a very low residual moisture content, and then heated to further increase its moisture retaining capacity. This hot dry air is then blown though an insulated hopper containing the pellets. The air both delivers heat to the pellets, and strips the moisture from their surface and carries it back to the dryer where it passes through a desiccant or a molecular sieve to remove the moisture. As the air rises through the hopper, it is cooled by contact with the material, progressively losing some of its drying capability, and so the airflow must be sufficient to heat the polymer throughout the hopper to the proper drying temperature. Thus there are four variables to be considered:

- Having the correct drying air temperature
- Achieving an adequate flow of dehumidified air through the pellets in the hopper
- Providing sufficient residence time for the pellets in the hopper to be properly dried
- Having dry air with the necessary moisture retaining capacity

Drying temperature

The proper drying temperature for PET is a function of the pellet residence time as shown in the graph. Typical drying hoppers are sized for 4 – 6 hours pellet residence time, resulting in a drying temperature for PET of between 155°C and 175°C measured at the air inlet to the dryer. Drying at lower temperatures will increase the required residence time. Sometimes PET is dried at higher temperatures than 175°C in order to reduce the heat input that is necessary during plasticizing and thereby shorten the plasticizing time and/or reduce the friction on the screw to minimise acetaldehyde (AA) formation. The maximum drying temperature should be limited to 190°C to avoid the possibility of bridging inside the hopper.

Typically PET pellets are dried at a temperature of 160°C for at least 4 hours. This temperature is considerably higher than the glass transition, T_g, of around 72°C at which temperature amorphous PET begins to soften and become tacky—making it essential that the PET pellets are crystalline, since material in the amorphous state would stick together, preventing the free flow of the pellets. It is also important that these times and temperatures are not exceeded, firstly because hydrolytic degradation is temperature dependent: it begins to occur in solid PET at temperatures as low as 150°C (although only very slowly) and the rate increases with temperature. Secondly, excessive drying times can lead to oxidative degradation (especially at higher temperatures), giving a yellow tint to the preform. Always turn down the dryer set temperature to 120°C should moulding be interrupted for more than a few minutes lest the recommended pellet residence time for the chosen drying temperature be exceeded and the material degraded. If a shutdown of more than 2 hours is anticipated, the dryer temperature can be reduced to 95°C in order to save energy.

Note that if the dryer contains multiple heating elements, the set point temperature may appear to be correct even though one heating element is defective. Monitoring the temperature of the material entering the feed throat of the machine will identify such a fault.

Airflow rate

Experiment has shown that for proper drying of PET the minimum flow of heated air required for each kg of PET processed per hour is 0.062 m^3/min. (For example, were 20 kg of PET to be processed each hour, then an air flow rate of 20 x 0.62 = 12.4 m^3/min would be required).

If the airflow is less than recommended, as a result of heat losses from the system the pellets near the top of the hopper are unlikely to reach the required temperature and so won't be adequately dried. The serious effect of having too low a drying temperature can be determined from the graph: at a pellet temperature of 178°C, PET takes only 4 hours to dry, but a 20% reduction in the temperature (i.e. to ~142°C) results in the pellets taking <u>almost three times</u> longer to dry.

An air filter is provided to prevent the desiccant material inside the dehumidifying dryer from becoming clogged with PET fines (dust) that would impair its efficiency. Regular cleaning of this filter is essential (the airflow through the dryer will be reduced if the filter is blocked).

Residence time

The design and capacity of the hopper must ensure all of the material is exposed to the hot dry air for the correct time. The bulk density of PET is typically around 0.84 kg/litre, although this does vary with pellet size in that smaller pellets will pack more closely together. To further complicate matters drying time is also affected by the size and shape of the pellets, since pellets with a high surface area to volume ratio enable the moisture to be removed more rapidly and therefore dry faster. Additionally the flow of pellets through the hopper is never true "plug-flow"; in fact the difference in residence time between pellets can be as much as 25%. To allow for all these different factors, a residence time of 6 hours is recommended. The residence time may be calculated by dividing the dryer capacity in kg of PET by the actual polymer throughput in kg/hour.

Moisture retaining capacity of the drying air

Heating of air produces a reduction in its relative humidity, but to achieve the far lower moisture content needed to prevent hydrolysis, dehumidifying dryers are needed. Inside a typical PET dryer air is dehumidified by being passed through a Silica Gel desiccant or a Molecular Sieve. Silica gel is an amorphous form of silicon dioxide produced in the form of hard irregular granules or beads. Its microporous structure of interlocking cavities gives it a very high surface area (800 square meters per gram) to which the water molecules adhere. Molecular sieves work in a different way to silica gel. They consist of highly porous crystalline metal-aluminosilicates that have many internal cavities linked by "window" openings of precise diameters. Only molecules with smaller diameters than these openings can be adsorbed. Larger molecules are excluded. This sieve-like selectivity, based on molecular size, plus a preference for polarizable molecules, makes molecular sieves ideal for adsorption of water, as water molecules are both polar and very small.

Molecular sieves differ from silica gel in a number of ways:

- They adsorb moisture more rapidly than silica gel
- They reduce water vapour to much lower levels than silica gel
- They perform more effectively as moisture adsorbers at higher temperatures than silica gel.

How dry does the drying air need to be? The amount of moisture (in grams of water per kg of dry air) that air at different temperatures can carry is plotted in a Mollier Diagram. The red line marks the saturation point—the minimum temperature at which the air can carry the given weight of moisture. If air at a certain temperature and pressure contains only half of the moisture that would saturate it under the given conditions, it is said to be 50% saturated. Connecting the 50% saturated points at each temperature gives the 50% relative humidity curve, and the same can be done with any percentage of the saturated moisture value. This diagram contains ten curves for relative humidity from 10% to 100% (saturation curve).

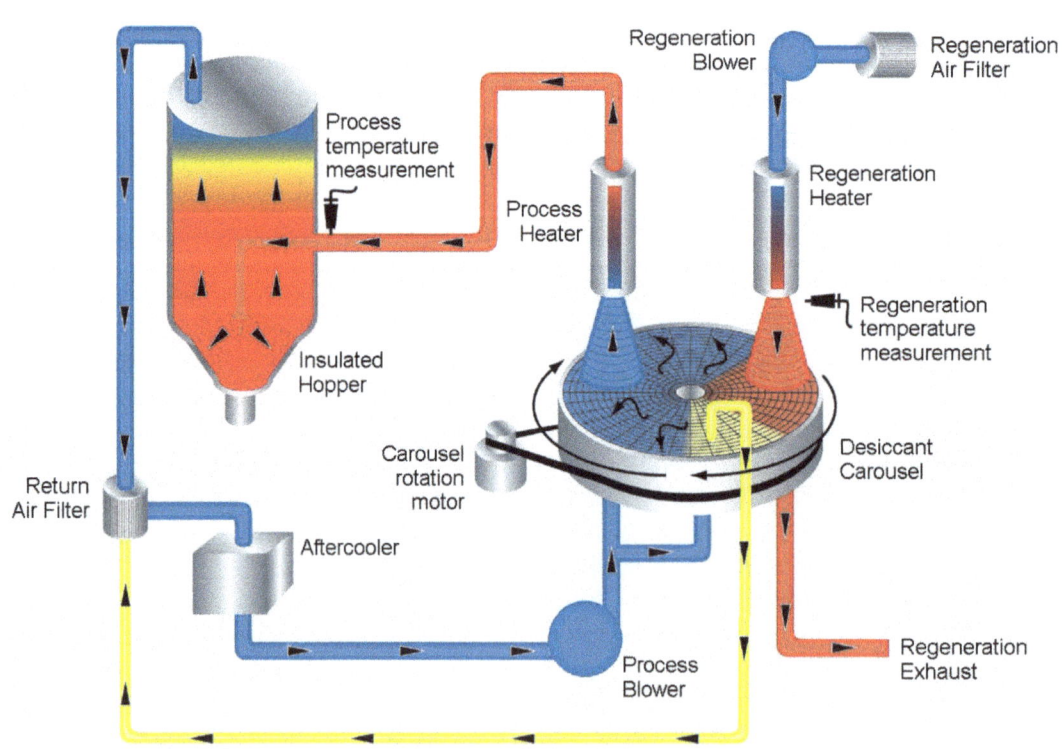

Schematic of a typical dehumidifying high temperature drying system

An example will show how the diagram is interpreted: identify the point representing an ambient temperature of 20°C and humidity of 35% RH, drop an imaginary line down from that point to the horizontal axis and read off the scale that each kg of air contains 5 grams of water. As air is cooled, its relative humidity will increase until it reaches 100%. This is the Dew Point where fog, mist, clouds or dew form. Dew point is a measure of the dryness of the air, and is expressed as a temperature in °C or °F. The lower its dew point temperature, the dryer the air. In the example, read horizontally across from where our imaginary line met the red saturation curve to find the dew point temperature of 3°C.

The Mollier Diagram not only makes it clear that the dew point of the air is directly related to its moisture content, but also highlights the dramatic increase in the amount of moisture that air is capable of carrying when heated, which helps explain why it is not necessary to go to extremes, and a dew point of "only" -20°C for the drying air is quite adequate, since a lower dew point will not materially improve dryer capability.

Most dehumidifying dryers employ a closed-loop drying-air circuit, since the "wet" air exiting the drying hopper is still drier than the atmospheric air, making it less costly to bring this air back down to the required dew point than it would be to dry ambient air.

Any desiccant or molecular sieve will have a finite moisture capacity, so periodically the adsorbed moisture must be purged. A *regeneration* circuit separate from the drying air circuit contains a blower which draws in ambient air through a filter and passes it through a set of heaters. This very hot regeneration air is then passed through the desiccant or molecular sieve which, as its temperature rises, releases the adsorbed moisture to be purged to the atmosphere. The hot, freshly regenerated desiccant/molecular sieve must be cooled (yellow circuit in the diagram) before being rotated back into the process-air drying circuit, to ensure that it remains effective in removing moisture from the drying air.

Many of the problems that occur in the production of PET bottles are directly related to low IV. So if there is a problem with the material distribution in the bottle, first confirm the dryer is functioning correctly by checking the drying air temperature, the air flow rate, and the dew point of the drying air. Other factors to consider with dehumidifying driers:

- The moisture-removing properties of the desiccant material will be lowered if the moist return air is too hot, so a heat exchanger is fitted to cool the moisture-laden return air before it passes through the desiccant. Regularly confirm that the heat exchanger is functioning as it should.

- Until it has cooled, hot desiccant will not absorb moisture very well. So if the drying air dew point is too high, it could be because the desiccant is too hot when it is rotated into the process air stream. Check that the regeneration system is functioning as it should.

- Dried PET at temperatures below 130°C re-absorbs moisture very rapidly when exposed to atmospheric air (which is very wet compared to the drying air) so that it is important to eliminate leaks. Also if the PET is dried away from the machine, it is essential that the dry pellets are conveyed to the machine using air which has a temperature and dew point similar to that of the drying air.

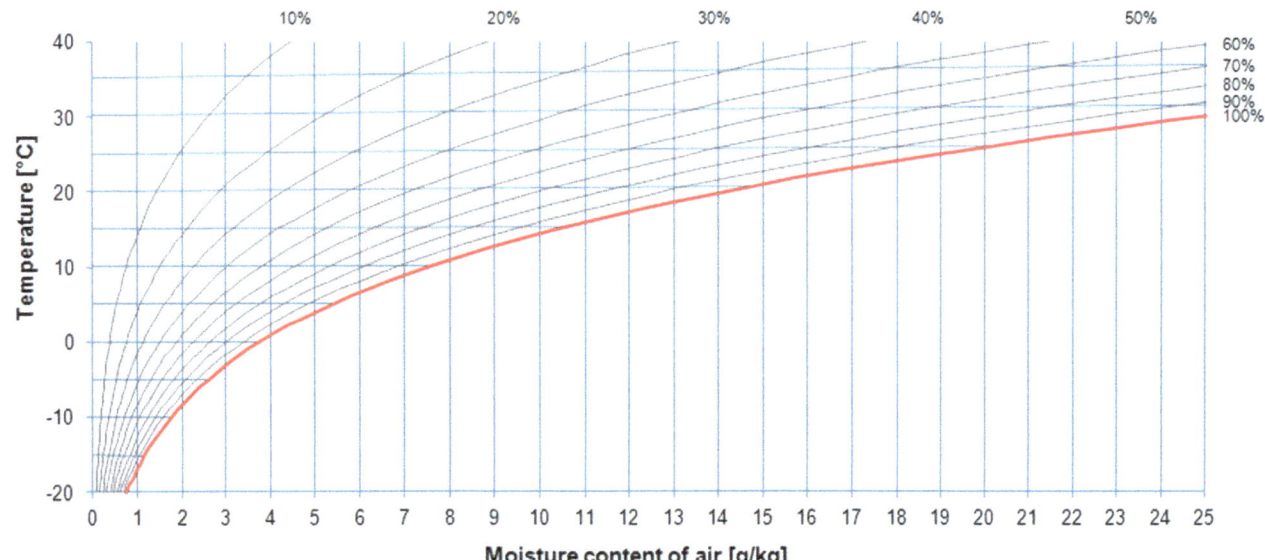

Mollier Diagram showing Relative Humidity curves at Standard atmospheric pressure

DRYING: AREAS TO BE CLOSELY MONITORED:

Assuming there are no obvious faults such as kinked or collapsed hoses, check:

1   Air Filter - Routine filter cleaning is essential. The filter prevents the desiccant material inside the dehumidifying dryer from becoming clogged with PET dust that might impair its efficiency.

2   Heat Exchanger - The moisture-removing properties of the desiccant material will be lowered if the moist return air is too hot, so a heat exchanger is normally fitted to PET dryers to cool the return air before it passes through the desiccant.

3   Air inlet temperature - Where the dehumidified air temperature is measured at a distance from the hopper inlet, the temperature setting may need to be raised to compensate for heat losses from the system and so achieve the required material temperature.

4   Ingress of ambient air - Dried PET re-absorbs moisture very rapidly when it is exposed to ambient air (which is very wet compared to the drying air) so that it is important to prevent any ingress that will reduce drying efficiency. If the PET is dried away from the machine, it is essential that the dried material be conveyed to the machine in dry air with a temperature and dew point similar to that of the drying air.

5   Drying Process Control - Most dryers monitor the drying air temperature; some also monitor the pressure drop across the filter (sounding an alarm if the filter becomes clogged) and the dew point of the drying air. Another useful check is to monitor the temperature of the material entering the feed throat of the machine. Any change in temperature provides an early indication of dryer malfunction.

Non-hygroscopic materials such as PP and PE pick up only surface moisture when stored in damp conditions. The drying of non-hygroscopic plastics does not require a high temperature dehumidifying dryer, but can be done in a simple hot air dryer.

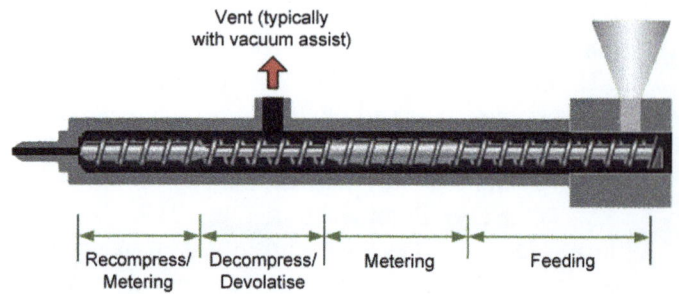

Vented injection barrel

Vented barrels have been advocated as a less-expensive alternative to using a dryer, or when reclaimed PET (which cannot readily be dried in a conventional dehumidifying dryer without first crystallizing it), is to be injection stretch-blow moulded. In this case a special two-stage screw is used in which the melt pressure is made to drop sharply in the area of the vent (to avoid melt leaking from the vent) and to build again afterwards. Any moisture present is released as vapour. Specific issues with this technique include preventing the PET from hydrolysing (i.e. reducing the IV) before the moisture is expelled, and keeping the vent clear. Vented barrels are not widely used with PET.

5

Hydraulics

Hydraulics is all about transmitting power by pushing on a confined incompressible fluid. An electrically driven pump produces fluid flow in a circuit consisting of valves, manifolds, and actuators joined together with steel pipe (with threaded joints), steel tube (with compression fittings or O-ring seals), or flexible hoses attached with reusable bolt-together end-fittings (used when the hydraulic lines are subject to flexing or vibration). Valves are either suitable for subplate mounting, or are modular in design and made with a standardized mounting surface enabling functional groupings to be built by stacking them together (so that no piping is required).

Important attributes of hydraulic actuators:

- speeds and applied forces can be varied
- motion can be instantly reversed
- the hydraulic fluid provides lubrication and cooling

When referring to the Aoki machine manual, note that hydraulic valves are prefixed with the letter V, and electrically energised solenoid operated directional valves are prefixed with the letters YV.

Hydraulic oil

Petroleum-based hydraulic oil is generally used within hydraulic systems because it transmits power readily, is only slightly compressible, lubricates the moving parts, seals small clearances, and cools or dissipates any heat that is generated. Particular care should be taken when selecting a hydraulic fluid that it is compatible with the hydraulic seals used in the valves and cylinders. It is also very important to monitor the quality and cleanliness of the hydraulic fluid by periodically having samples analysed (and the result recorded). When the contamination level reaches a set level, the oil should be replaced.

Oil tank

The hydraulic oil is contained in a reservoir situated beneath the electric motor and hydraulic pumps. The oil tank incorporates several important features: a filler/breather to maintain atmospheric pressure within the tank by accommodating the air exchange that occurs as the hydraulic cylinders extend and retract (and which needs to be cleaned monthly); a sight glass which allows the fluid level to be checked visually without risking contamination (in some machines a float switch is also fitted); clean-out plates installed in the side(s) of the tank; an oil temperature sensor; and a large magnet placed inside to attract any ferrous metal contamination that may be suspended in the oil.

Oil is drawn from the tank by the pump through a submerged suction filter, and returns (after first passing through the heat exchanger) to the opposite side of the tank via a return line set below oil level to avoid aeration. Because oil returning to the reservoir

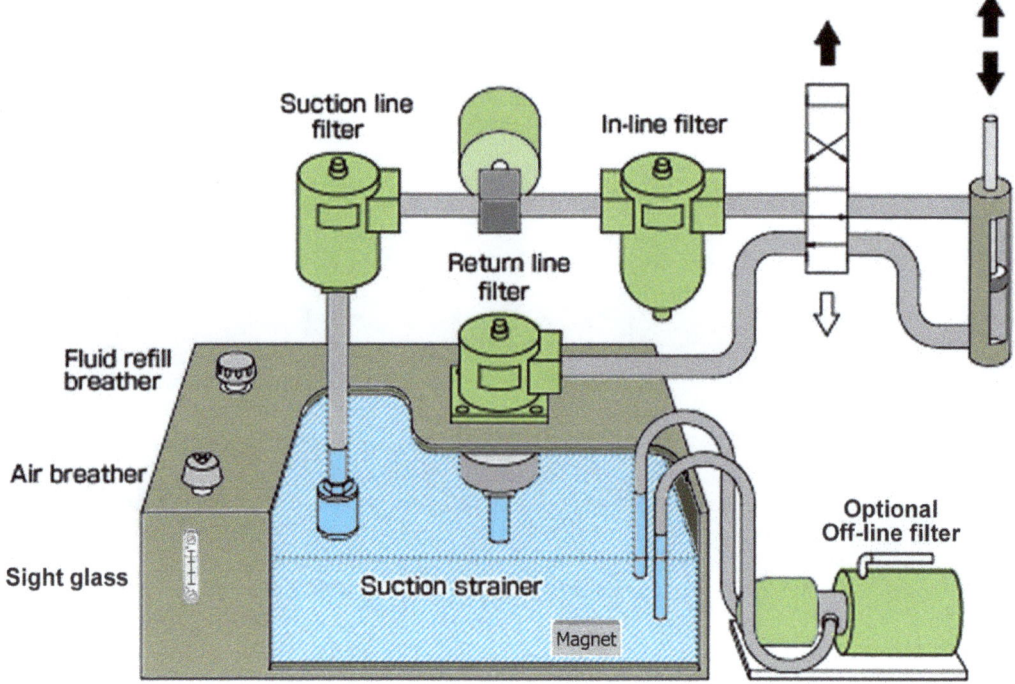

is usually warmer than the supply oil and frequently contains air bubbles, a baffle is provided to prevent the returning oil from directly re-entering the pump inlet. This forces it to move along the walls, where much of the heater is dissipated to the walls of the reservoir. It also allows the contaminants to settle at the bottom of the tank, and provides an opportunity for any entrained air bubbles to clear.

Filters

In order to trap particulate contamination of the oil, suction filters are fitted at the inlets to the pumps. They are fitted with a relief valve which opens when the filter becomes too clogged to pass full flow. Opening of the bypass is shown either by an alarm sounding, or by a green indicator at the top of the filter changing to red. (Note that when the oil temperature is low the red indicator may also appear, but if it can be reset, and it remains green once the 45°C operating temperature has been reached, then the filter is still in a usable condition). An additional filter is fitted in the supply line to the Rotary Actuator. The reusable filter elements are fabricated from stainless steel mesh which must be cleaned using ultrasonic cleaning techniques.

Heat Exchanger

A heat exchanger prevents the hydraulic oil from getting too hot. It contains multiple cooling tubes through which the oil is fed at low pressure before being returned to the tank. Surrounding the cooling tubes is a jacket through which cooling water is fed. As the oil passes along the tubes, its heat is transferred to the cooling water. To ensure more stable running conditions, the hydraulic oil temperature can be more closely controlled by using an optional thermostatic valve to regulate the flow of cooling water through the heat exchanger. The working temperature of the oil should be 45°C with the maximum oil temperature not exceeding 50°C.

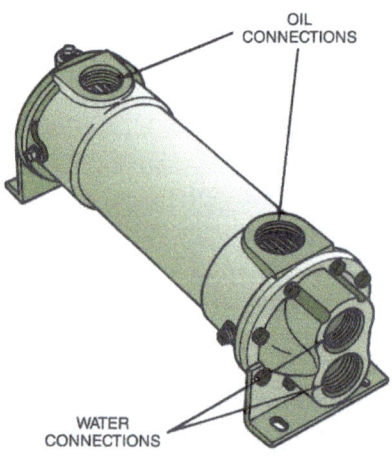

It is essential that cooling water never gets into the hydraulic oil (readily identifiable by the distinctive milky appearance of the resultant mixture). To prevent the steel rusting, sacrificial zinc anodes are fitted inside the heat exchanger. Since zinc oxidizes more easily than iron, it corrodes first. The sacrificial anodes should be periodically checked and cleaned to ensure they remain in good condition. Once the zinc is mostly consumed, the anodes need to be replaced.

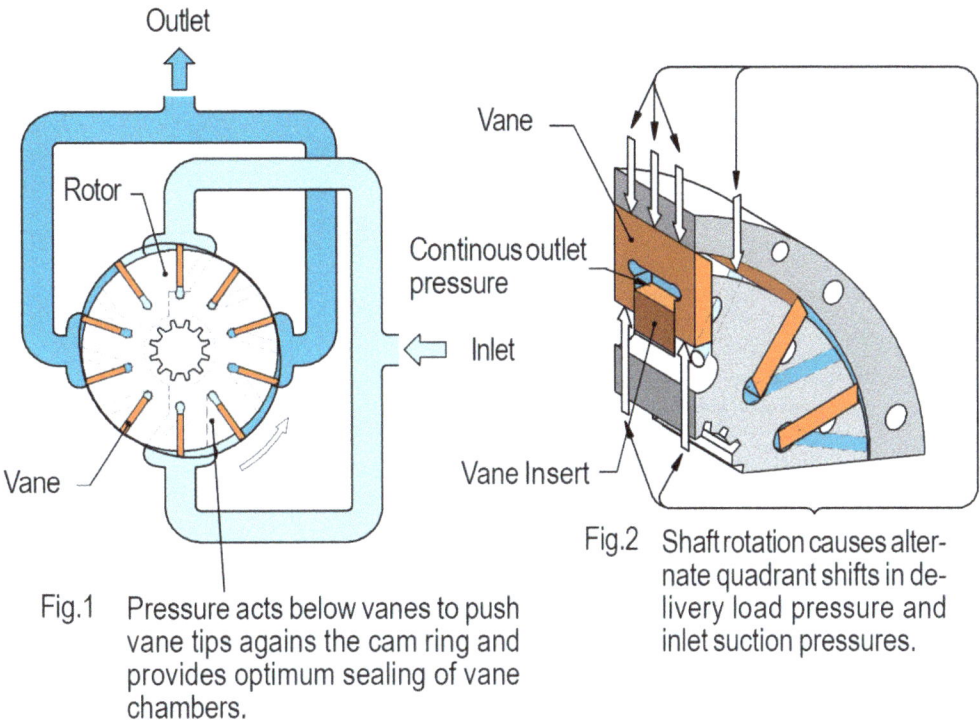

Fig.1 Pressure acts below vanes to push vane tips agains the cam ring and provides optimum sealing of vane chambers.

Fig.2 Shaft rotation causes alternate quadrant shifts in delivery load pressure and inlet suction pressures.

At one time cooling water was typically supplied from evaporative cooling towers because they were relatively inexpensive in terms of capital and energy costs, but now that the law requires regular biocide treatments to prevent the growth of legionella bacteria, closed circuit air blast systems (i.e. comparable to a large car radiator) are preferred. One possible disadvantage with such systems is that the minimum temperature that can be achieved is only 5°C below ambient (and accordingly varies with the ambient temperature). However in all but the warmest climes this is adequate for cooling the hydraulics on moulding machines (although refrigeration systems are required for chilling injection moulds).

Hydraulic pump

All classic one-stage machines are fitted with combination vane pumps as shown in the diagram. A slotted rotor splined to the drive shaft turns inside an elliptical cam ring. Vanes, which slide in slots positioned around the rotor, are pressed outwards against the inside of the cam ring by the oil pressure generated as the rotor turns, thus creating multiple pumping "chambers" between themselves, the rotor on the inside, the cam ring on the outside, and the two side plates. Two pairs of opposed inlet and outlet ports are interconnected via passages within the housing. Because the ports are positioned 180 degrees apart, forces caused by pressure build-up on one side are cancelled out by equal but opposite forces on the other, preventing side loading of the drive shaft and bearings.

Because the cam ring is elliptical, as the rotor turns the 'chambers' initially increase in size, creating a partial vacuum that collects fluid entering the inlet port, and then become progressively smaller, forcing the fluid to be expelled at the pump outlet. The displacement of the pump is not adjustable, depending as it does on the widths of the ring and rotor, and on the distance the vanes extend from the rotor surface to the ring surface. Typically there are three or sometimes even four pumps driven by a single Electric Motor:

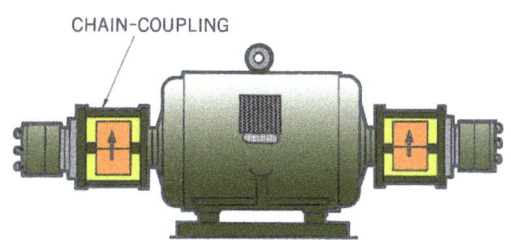

Chain-couplings are used to connect the shafts of the electric motor and the hydraulic pumps because they are capable of transmitting relatively high torques in a minimum of space, while accommodating limited misalignment. (They require less space than other methods because they apportion the torque over the entire length of roller chain and all around the sprocket teeth). The split covers which protect the chain and sprockets from contamination need to be removed every six months so that the chains and sprockets can be inspected and re-greased. It is important not only to use the correct grade of grease, but also to be aware that because of centrifugal force, there is a tendency for the grease to stick to the inside of the covers, resulting in poor lubrication of the chain and sprockets.

In the hydraulic diagram below oil from the pump is fed via a directional valve to both ends of the cylinder. A *large volume* of oil is required to extend the cylinder at high speed, but once it has fully extended the flow stops and *high pressure* is needed to apply maximum force. These two apparently conflicting requirements can be resolved by having two separate pumps known as a combination pump. One pump delivers "high volume at low pressure", and the other "high pressure but low volume".

Note that hydraulic pumps require a *Pressure Relief Valve (#8)* to control and limit the pressure generated in the system to prevent catastrophic failure.

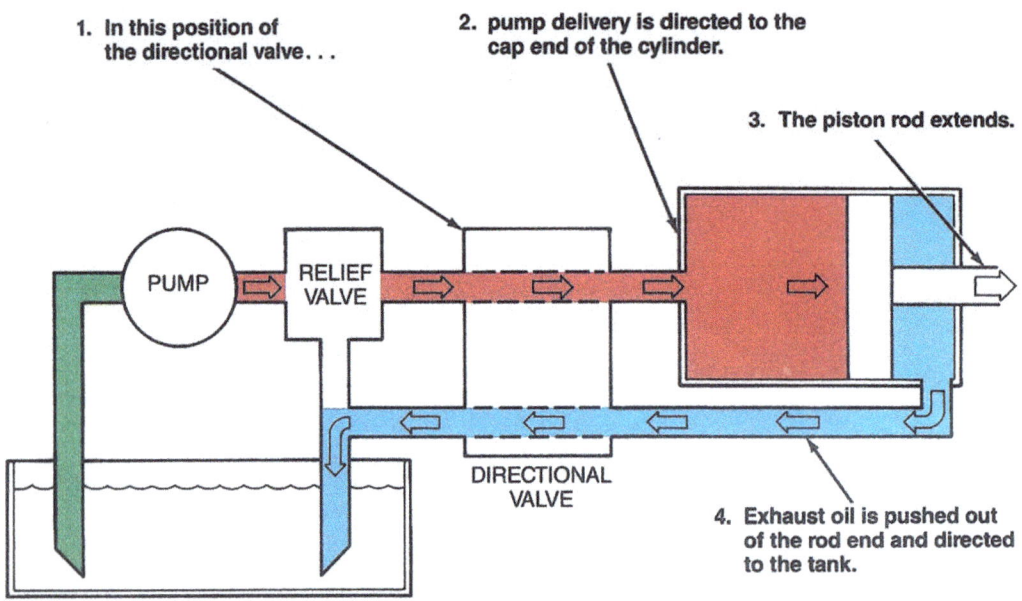

1. In this position of the directional valve...
2. pump delivery is directed to the cap end of the cylinder.
3. The piston rod extends.
4. Exhaust oil is pushed out of the rod end and directed to the tank.

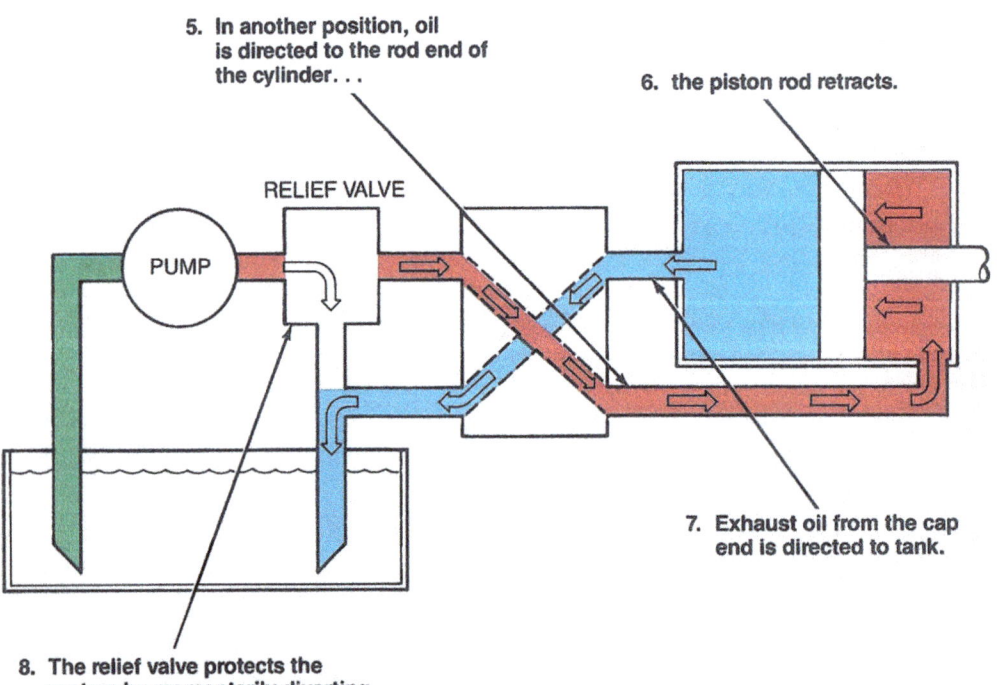

5. In another position, oil is directed to the rod end of the cylinder...
6. the piston rod retracts.
7. Exhaust oil from the cap end is directed to tank.
8. The relief valve protects the system by momentarily diverting flow to tank during reversing, and when piston is stalled or stops at end of stroke.

Chapter 5—Hydraulics 51

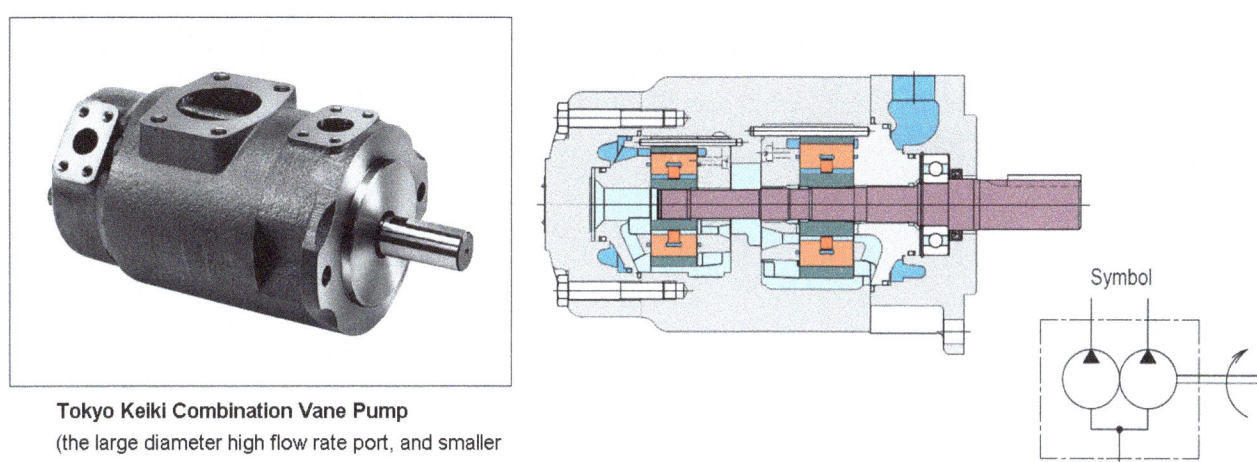

Tokyo Keiki Combination Vane Pump
(the large diameter high flow rate port, and smaller diameter high pressure port are clearly visible

Pressure Relief Valve

A normally closed valve (#8 in opposite diagram) connecting the pressure line to the oil tank. Pressure Relief valves protect the hydraulic system from excessive pressure, and are also used to maintain constant pressure in a hydraulic system. They work by diverting some or all of the flow to tank when the set pressure is reached (see diagrammatic explanation below).

Pilot-operated Relief Valves

A pilot-operated relief valve operates in two stages. The pilot stage in the upper valve body contains the pressure limiter: a poppet held against a seat by an adjustable spring. The port connections are made to the lower body, and diversion of the full flow volume is accomplished by the lifting of the balanced piston in the lower body.

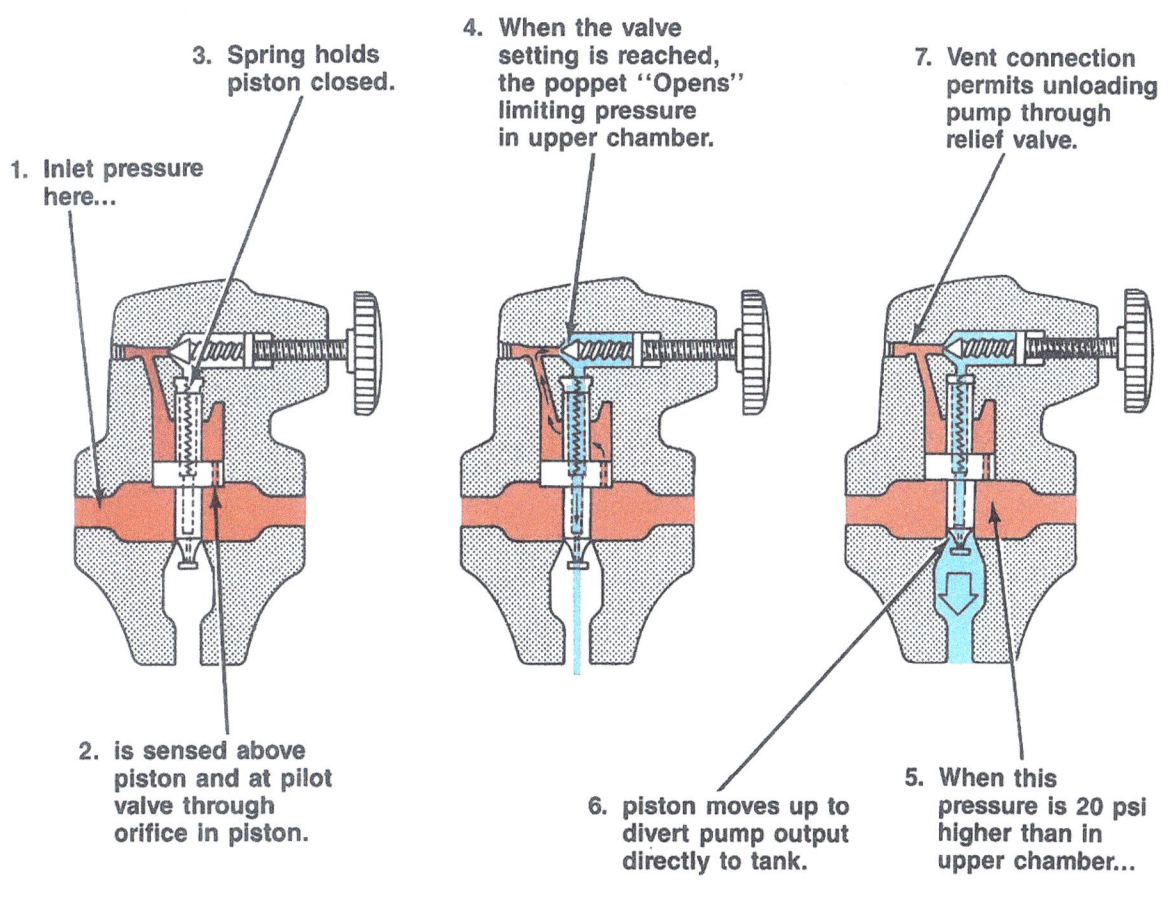

1. Inlet pressure here...
2. is sensed above piston and at pilot valve through orifice in piston.
3. Spring holds piston closed.
4. When the valve setting is reached, the poppet "Opens" limiting pressure in upper chamber.
5. When this pressure is 20 psi higher than in upper chamber...
6. piston moves up to divert pump output directly to tank.
7. Vent connection permits unloading pump through relief valve.

A. CLOSED   B. CRACKED   C. RELIEVING

In normal operation the balanced piston is in hydraulic balance (hence its name). Pressure at the inlet port acting under the piston is also felt on its top via a small orifice drilled through the piston. At any pressure less than the valve setting (NB screw *in* to increase set pressure), the piston is kept on its seat by a light spring, but when the pressure reaches the set value, the poppet is forced off its seat, thereby connecting (via the channel through the hollow stem) the upper chamber to tank. The restricted flow through the orifice is unable to replenish the upper chamber quickly enough, and so once the difference in pressure between the upper and lower chambers is sufficient to overcome the force of the light spring (approximately 0.4 MPa), the large piston unseats, permitting full flow directly to tank.

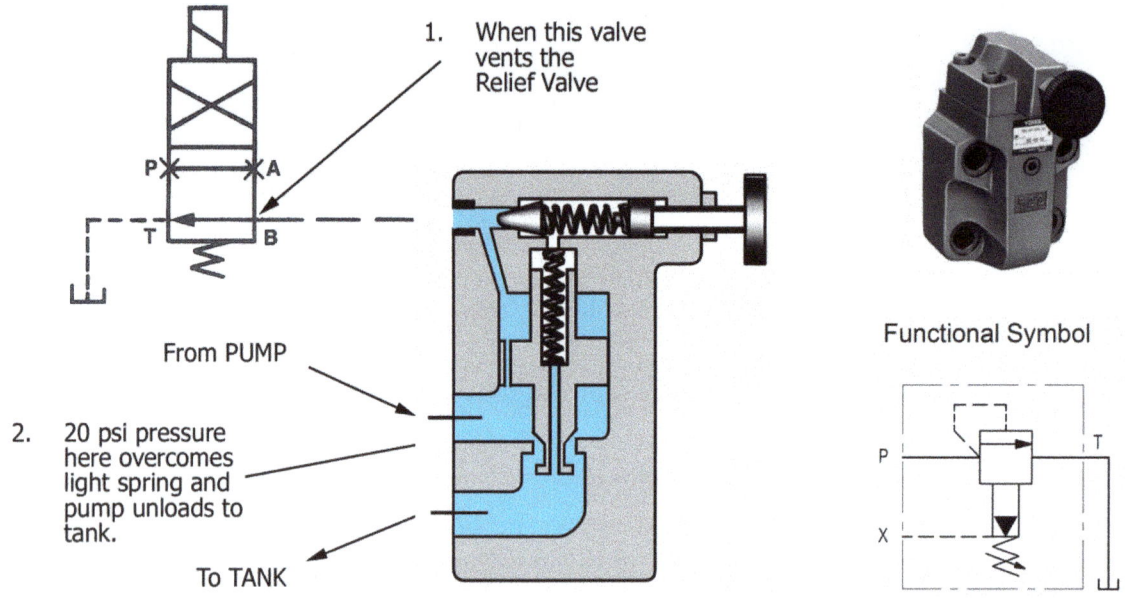

A relief valve may be "remotely" controlled by means of the *vent* port connected to the upper chamber. When this chamber is vented to tank, the only force holding the piston on its seat is that of the fore mentioned light spring, and so the valve will open fully at approximately 0.4 MPa. In order to reduce hydraulic shock as the vent port is opened, a *shockless* valve may be sandwiched between the controlling solenoid-operated directional valve and the vent port (not shown).

Using the same port, a relief valve may also be fitted with a pilot operated "override" so that one valve can relieve at two different set pressures. To exercise control, the second valve must be set at a lower pressure than the integral poppet.

A remotely controlled relief valve combined with a solenoid operated directional valve as in the circuit opposite, may be used to control *three* different pressures. In this case when the solenoid valve is NOT energised, the pressure in the upper chamber is vented to tank and so the pressure in the system is at a minimum. When solenoid-b is energised, remote control valve ® controls the pressure in the system, whereas energising solenoid-a allows the setting on the balanced piston relief valve itself to control the system pressure.

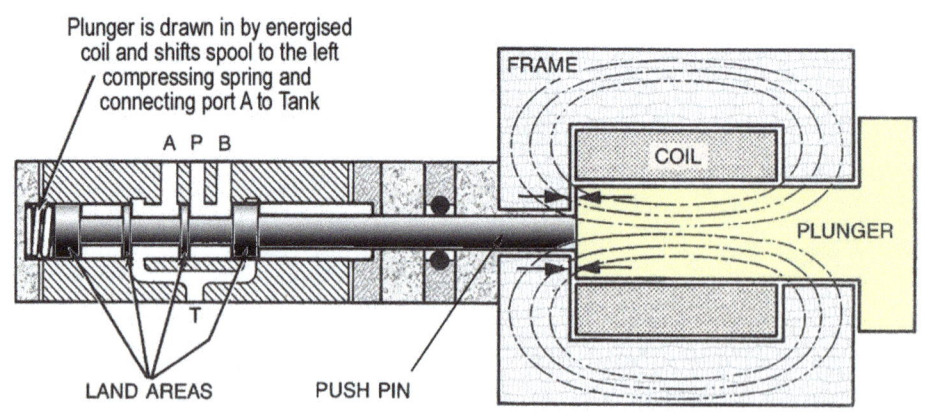

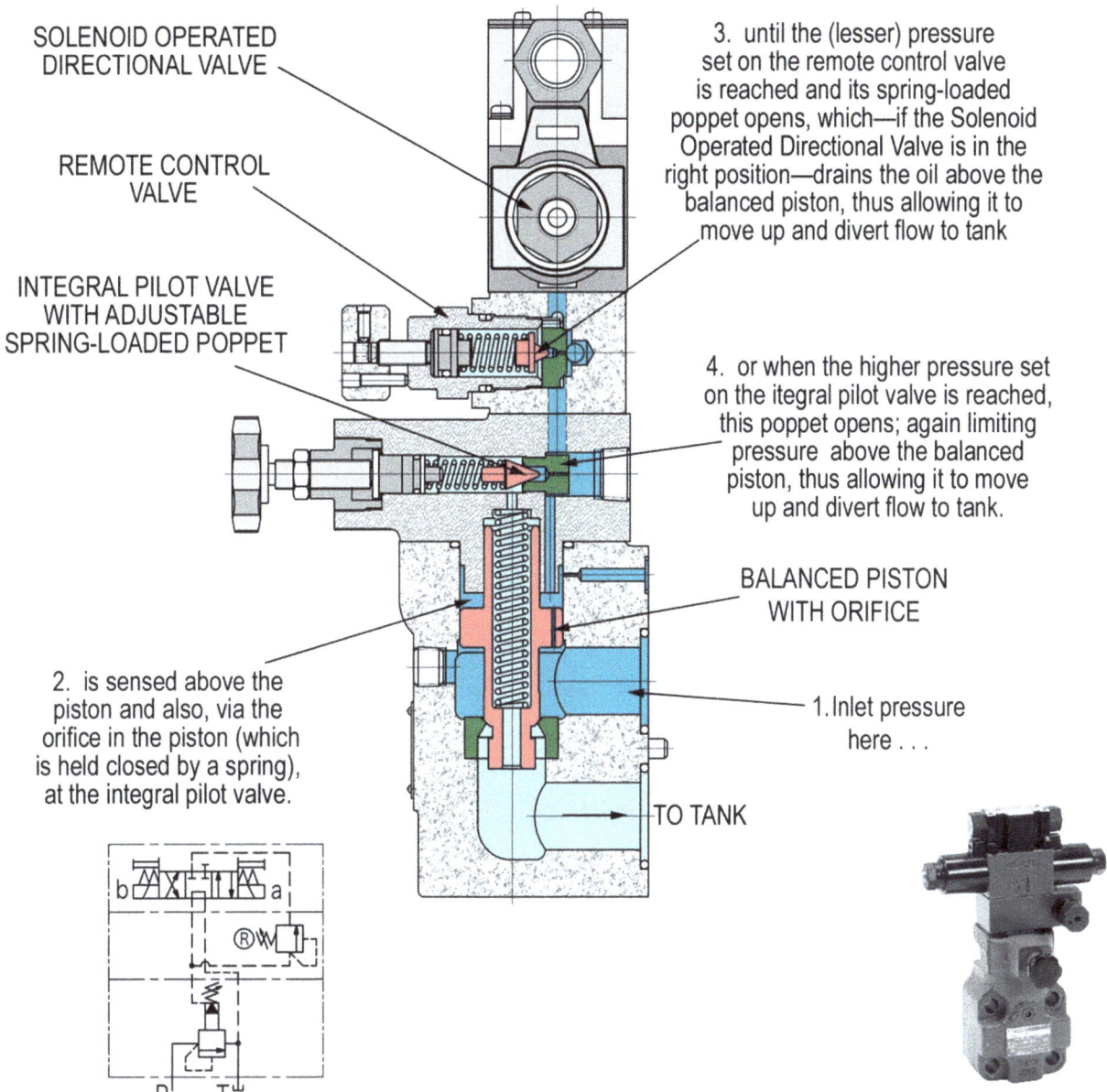

Directional Valve

As the name implies, directional valves start, stop, and control the direction of fluid flow. Although they share this common function, directional valves vary considerably in construction and operation. Typically they are subplate mounted and the functional element is a sliding spool which is shifted either electrically or hydraulically.

Electrical actuators are commonly called solenoids. Electric current flowing through a conductor produces a magnetic field. If the conductor is formed into a coil the strength of the magnetic field will increase. As long as current is flowing through the wire, the coil acts like an electric magnet. The strength of the magnetic action can be increased by placing a soft iron core inside the centre of the coil. If the iron core is allowed to move, it will centre itself within the coil when the current flow produces a magnetic action. This centring action provides the linear mechanical motion required for solenoid operation. The soft iron core is mechanically connected to the directional valve spool so that when current flows in the coil, the resulting magnetic field moves the core or *plunger*, and thus the spool.

The operation provided by a Directional Valve is dependent on the spool and body configuration. Most three-position spool valves provide the same flow paths when the spool is shifted, but various centre conditions are available (which means that should a valve have to be replaced, great care must be taken that the replacement is identical in operation to the original).

A direct-acting solenoid valve for controlling large volumes of fluid would itself be massive, and thus require a significant amount of electricity to shift its spool. Instead *two-stage* valves have been developed which use an electrically-operated pilot stage to hydraulically shift the main spool by directing fluid to one end or the other while connecting the opposite end to the tank.

To control the shift speed of the main spool, a flow restrictor in the form of a fixed orifice can be fitted between the main stage and the pilot valve. This will provide smoother reversals and prevent hydraulic shock. (NB: Should it be necessary to replace one of these valves, it is essential that the correct orifice diameter is fitted). To operate the valve manually, push in the *manual override* located in the centre of the solenoid assembly (note that the manual actuation force is greater the higher the system pressure).

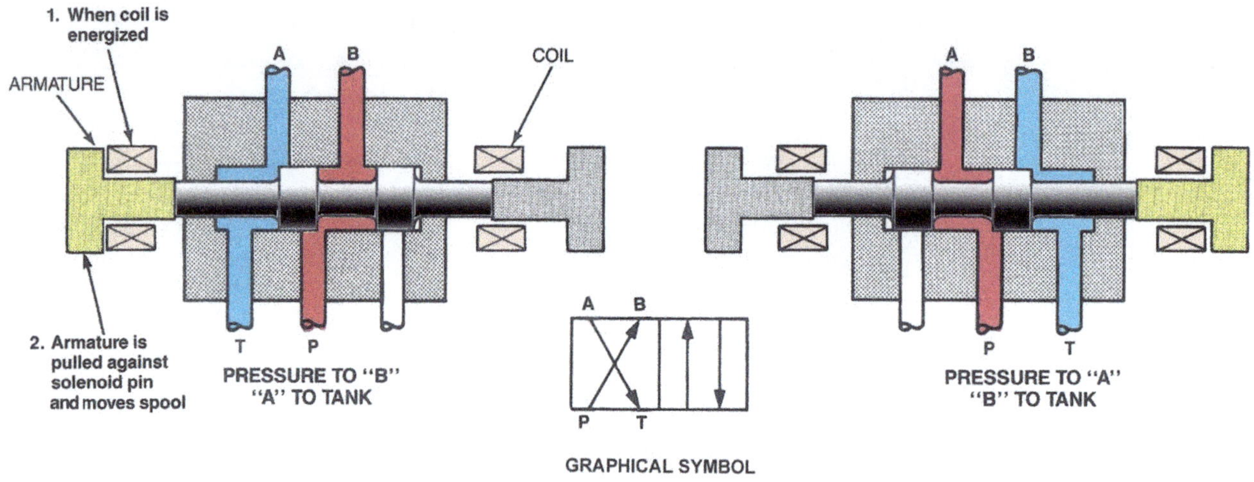

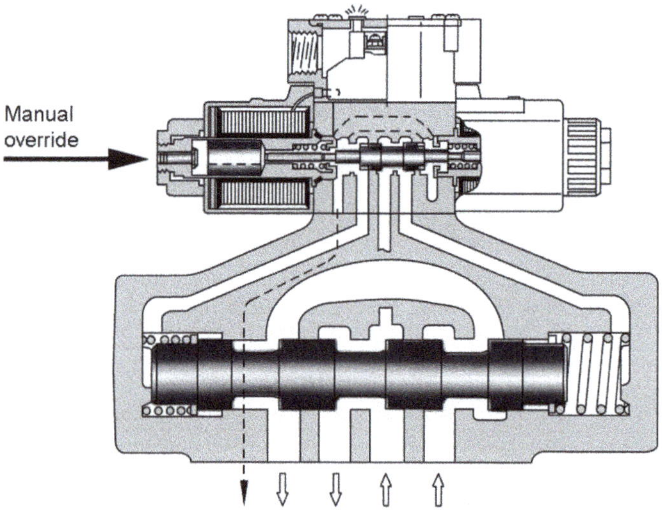

Pressure Reducing valve

A valve used to reduce the pressure in a part of the circuit to the level required by certain components. A spring loaded spool limits the downstream pressure. A drilling in the spool allows the oil pressure to be felt against the end of the spool. Should this pressure be high enough to overcome the spring force, the spool will shift to allow enough hydraulic oil to drain back to tank to prevent the pressure rising any further.

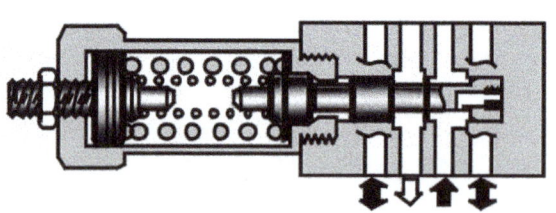

Relief Valve used as an unloading valve

Hydraulic circuit (A.) below illustrates the arrangement of components in an *unloading* system with the flow condition at low pressure.

Oil from the larger volume pump passes through the unloading valve (C) and over the check valve (D) and combines with the output of the low volume pump.

This situation continues as long as the system pressure is lower than the setting of the unloading valve.

In circuit (B.), system pressure is equal to the setting of the unloading valve which accordingly opens, permitting the larger volume pump to discharge to the tank at little or no pressure. The check valve (D) closes, preventing flow from the pressure line through the unloading valve (C). In this condition, much less energy is used than if both pumps have to be driven at high pressure. However the final advance will be slower because of the smaller volume output to the system. When motion finally stops, the small volume pump discharges through the relief valve (E) at its set pressure.

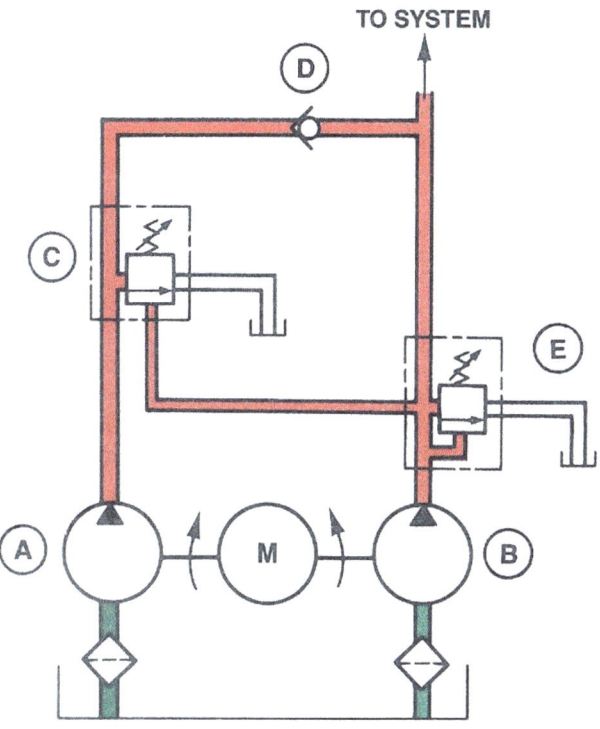

A. LOW PRESSURE OPERATION

System pressure is less than the adjusted settings of pressure control valves (C) and (E). Therefore, both (C) and (E) are in their normally closed positions. Delivery of pump (B) is directed into the system through (E). Delivery of pump (A) is directed through (C) and check valve (D) and combines with delivery of (B) to also be directed into the system.

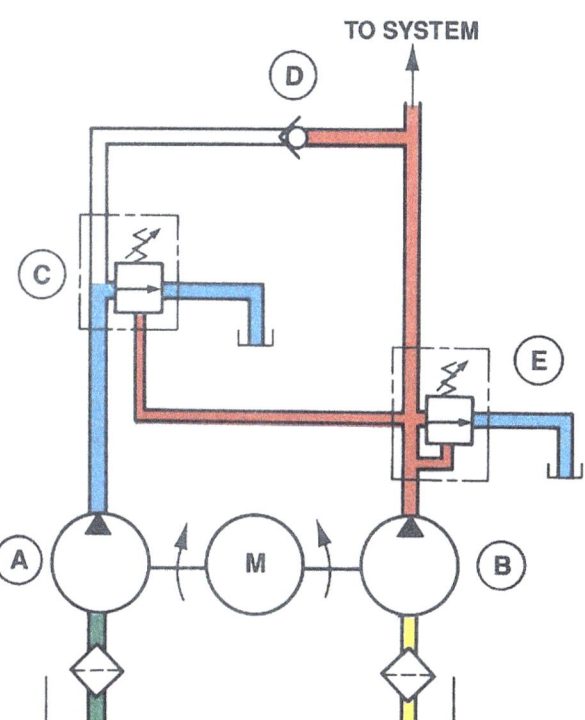

B. HIGH PRESSURE OPERATION

System pressure is equal to the setting of relief valve (E) but higher than the setting of unloading valve (C). Delivery of pump (B) returns to the tank through valve (E) at its pressure setting. Delivery of pump (A) returns to the tank through valve (C) at a very low pressure (unloaded) as (C) is held wide open by system pressure.

Check Valve

Allows free flow in one direction, while blocking flow in the other direction. It may be manually, solenoid, or pilot actuated with spring return (often two or more methods are combined). As with the standard Check Valve, a "pilot-to-open" Check Valve permits free flow in one direction and prevents flow in the reverse direction, but when a pilot pressure signal is applied to the pilot port, free reverse flow is permitted. Shuttle (selector) Valves allow flow from two different directions to exit through a single outlet.

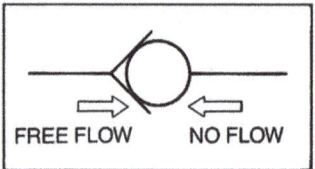

Logic (Cartridge) valves

Development of hydraulic systems has led to an increased use of manifold blocks and slip-in *Logic* or *Cartridge* valves. Slip-in Cartridge Valves are rather like poppet check valves and consist of an insert assembly that slips into a standard cavity machined into a manifold block. A Control Cover bolted to the manifold secures the insert within the cavity, eliminating potential leakage points. Cartridge Valves offer a design alternative rather than a replacement for conventional sliding spool valves. Often, the most economical system employs a combination of Cartridge Valves and conventional sliding spool valves mounted on a common manifold.

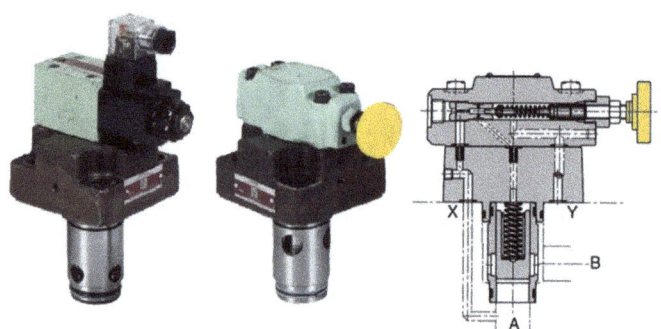

Flow Control Valve (throttle valve)

The oil flow created by the pump gives an actuator its motion (pressure gives the actuator its force, but flow is essential to cause movement). By controlling the rate of flow, it is possible to control the speed of hydraulic cylinders or motors. In its simplest form, the Flow Control Valve is nothing more than a variable orifice. By varying the size of the opening, the amount of oil entering a cylinder can be varied and thus the piston speed changed. Frequently Flow Control Valves are integrated with Check Valves as shown in this modular valve:

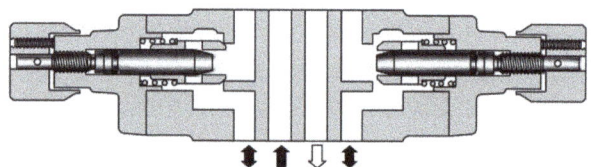

Regenerative Circuit

Rapid extension of a hydraulic cylinder rod can be achieved by returning the flow of oil from a cylinder's rod end back into its cap end. Termed "regeneration", the principle is illustrated in the circuit opposite. Note that the "B" port on the direction valve, which would normally connect to the cylinder, is plugged and the rod-end of the cylinder is connected directly to the pressure line. With the valve shifted to connect the "P" port to the cap-end, flow out of the rod-end joins pump delivery, increasing the cylinder speed. In the reverse condition, flow from the pump is directed to the rod-end. Exhaust flow from the cap-end returns to the tank through the direction valve.

Note also that the pressures on the cap side and rod side of a cylinder in a regenerative circuit are effectively equal. What is causing the cylinder to extend is the difference in force between the cap end and the rod end due to the difference in their respective areas.

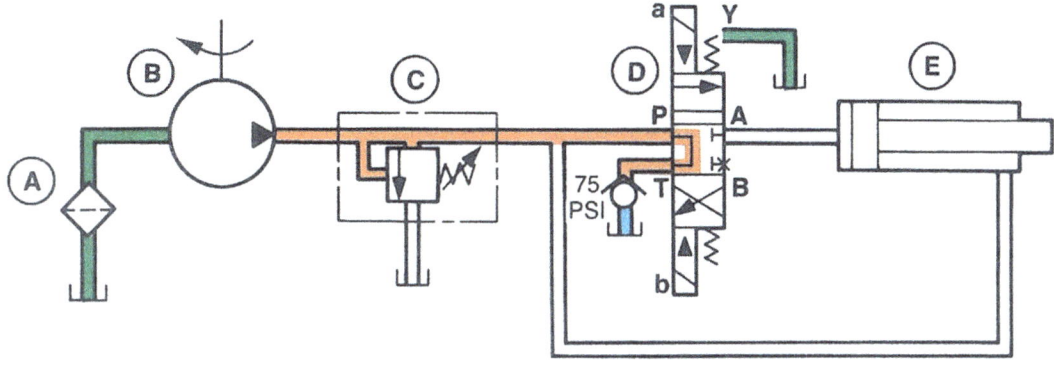

A. IDLE

View A shows the idle condition of the circuit when solenoids (Da) and (Db) are both de-energized. The pump delivery is unloaded through valve (D) and a 75 psi check valve.

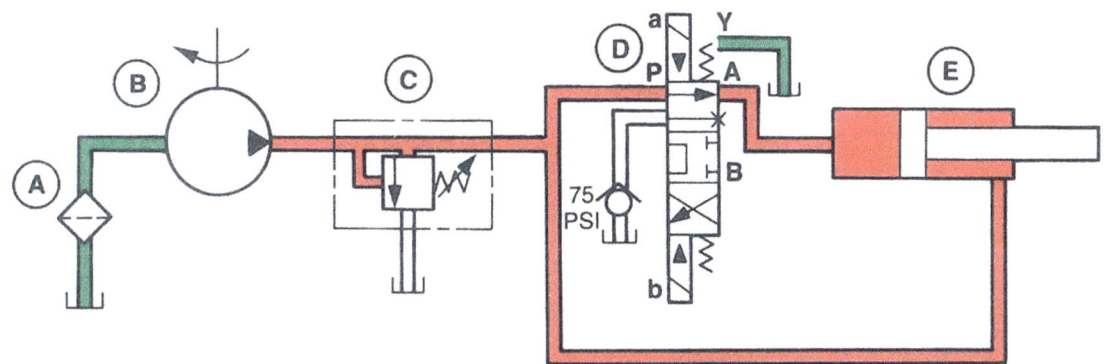

B. ADVANCING

View B shows the flow and force conditions when solenoid (Da) is energized for regenerative advance. Discharge from the rod end of the cylinder joins the pump delivery at the "P" port of valve (D) to increase the piston speed. However, note that system pressure also acts in the rod end of the cylinder to reduce the output force capability.

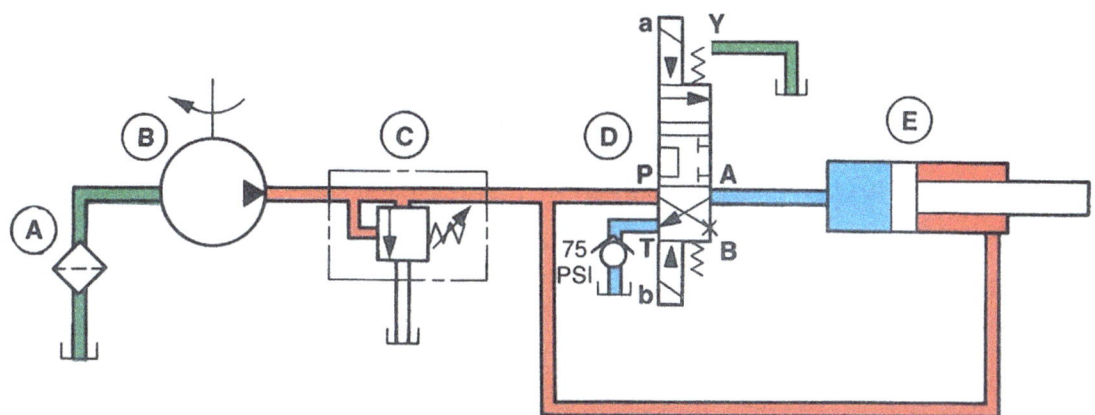

C. RETRACTING

View C shows the flow and force conditions when solenoid (Db) is energized. The pump delivery is directed into the rod end of the cylinder. The cap end of the cylinder is returned to the tank via "A" to "T" through valve (D) and a backpressure check valve.

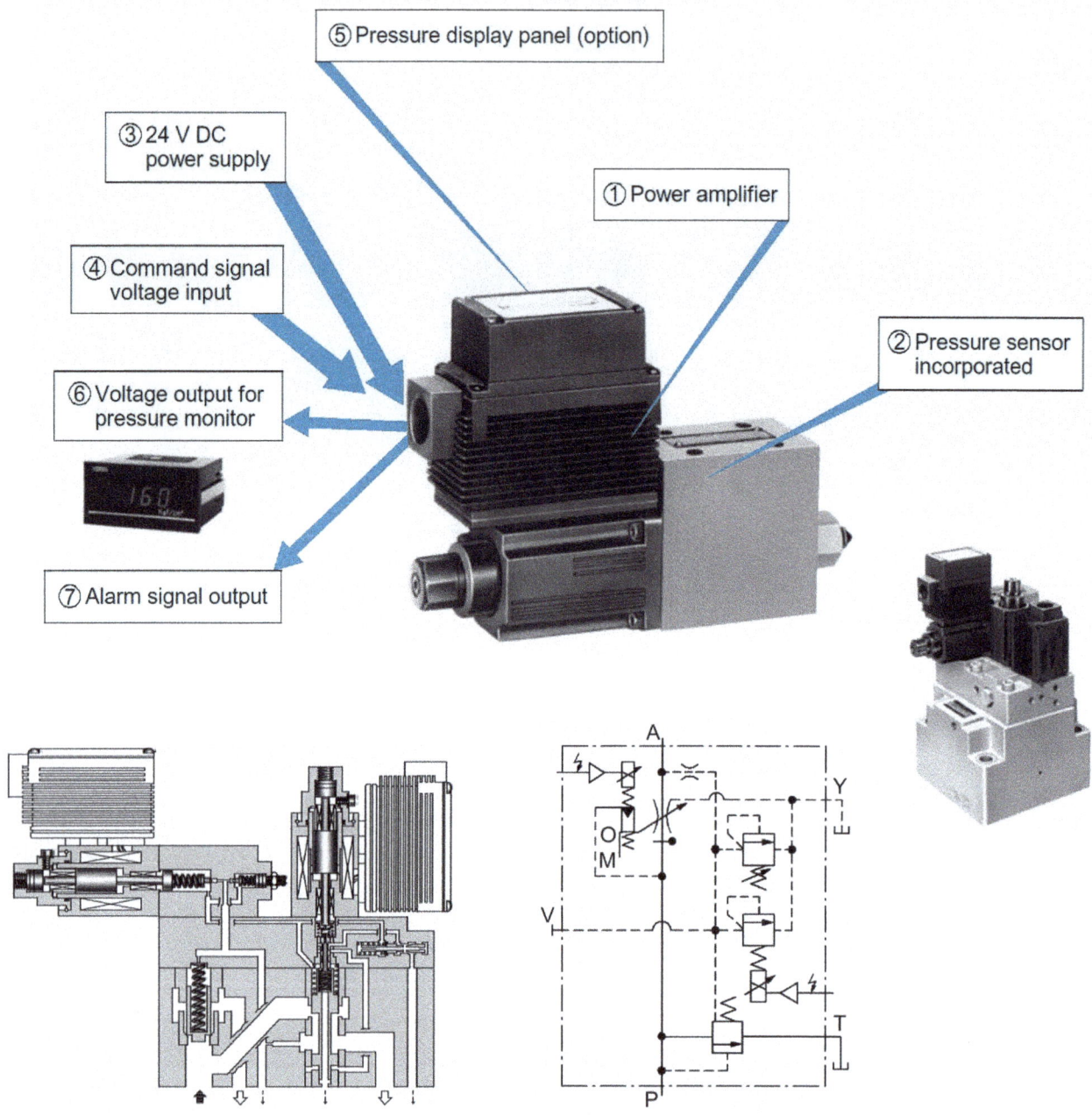

Proportional valves

Proportional valves deliver smooth "step less" motion by constantly adjusting pressure and flow in response to feed back of the actual pressure in the system and the precise position of the actuator. They are solenoid operated spool valves in which the solenoid current can be varied to make the spool move variable distances. Since proportional valves supply the minimum pressure and flow required to operate actuators they are often termed 'energy saving' valves.

In machines fitted with an injection controller, the injection ram position and pressure are monitored by the Injection Controller which then varies the electrical signal to the solenoid in order to achieve the required injection speed and pressure.

Oil motor

Either a fixed displacement *in-line* (axial) or *radial piston* oil motor provides rotary motion for the injection screw. With the in-line piston design overleaf, the motor drive shaft and cylinder block are centred on the same axis. Pressure at the piston ends causes a reaction against a swash plate, driving the cylinder block and motor shaft in rotation.

With the radial piston "Staffa" motor pictured right, oil from the pump is initially supplied to the distributor valve spool of the motor. Four pressure balancing slots, two on each side of the spool, offset the side load due to the pressure forces acting on the pressure distributor ports of the valve; thus keeping the valve centred with no frictional or efficiency losses. The distributor valve directs the oil to two of the cylinders at a time so that the pistons in these cylinders push down on the off-centre throw of the crankshaft (drum), causing it to rotate. At the same time the distributor valve enables the remaining cylinders to exhaust their oil back to the exhaust port. The distributor valve spool is keyed to the crankshaft so that both rotate together and port oil to each of the cylinders in succession (the timing is slightly out of phase between the two so there is no "dead spot" at top dead centre).

Piston forces are transmitted to the shaft by connecting rods through a large contact area, self-aligning ball and socket joint. Oil is supplied under pressure via a hole in each piston and connecting rod to the slipper bearing surface. This hydraulically balances the piston-connecting rod assembly so that it rides lightly on the crankshaft. As all moving parts are immersed in hydraulic oil, "Staffa" motors do not require periodic lubrication.

The *torque rating* [Nm] of an oil motor represents the maximum amount of turning moment it can produce at the specified hydraulic pressure. A large diameter injection screw will require a higher torque to turn it than will a small diameter screw. A viscous material such as PET needs a high torque coupled with a relatively low rotary speed to avoid excessive shear-induced heating.

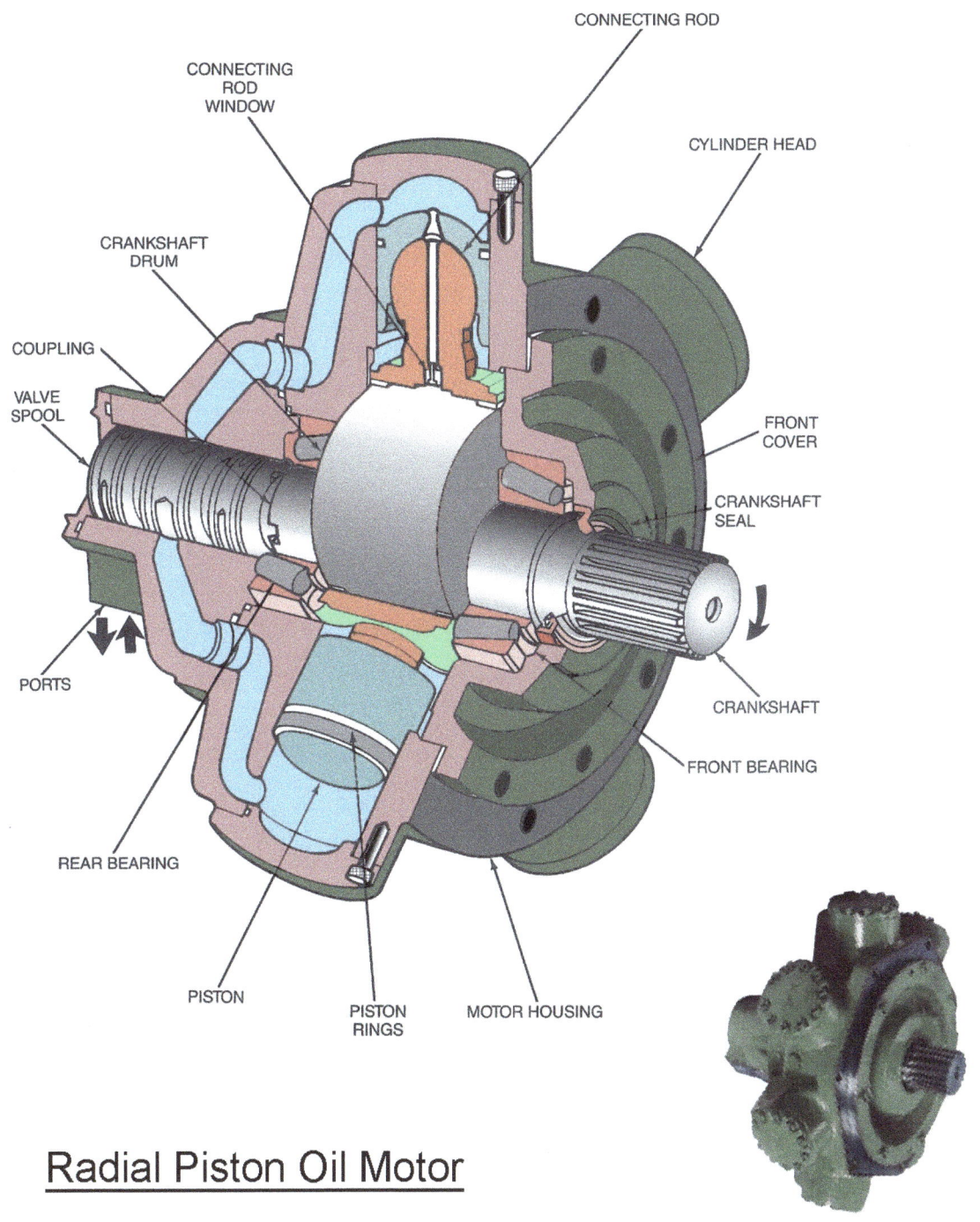

Radial Piston Oil Motor

In-line (axial) Oil Motor

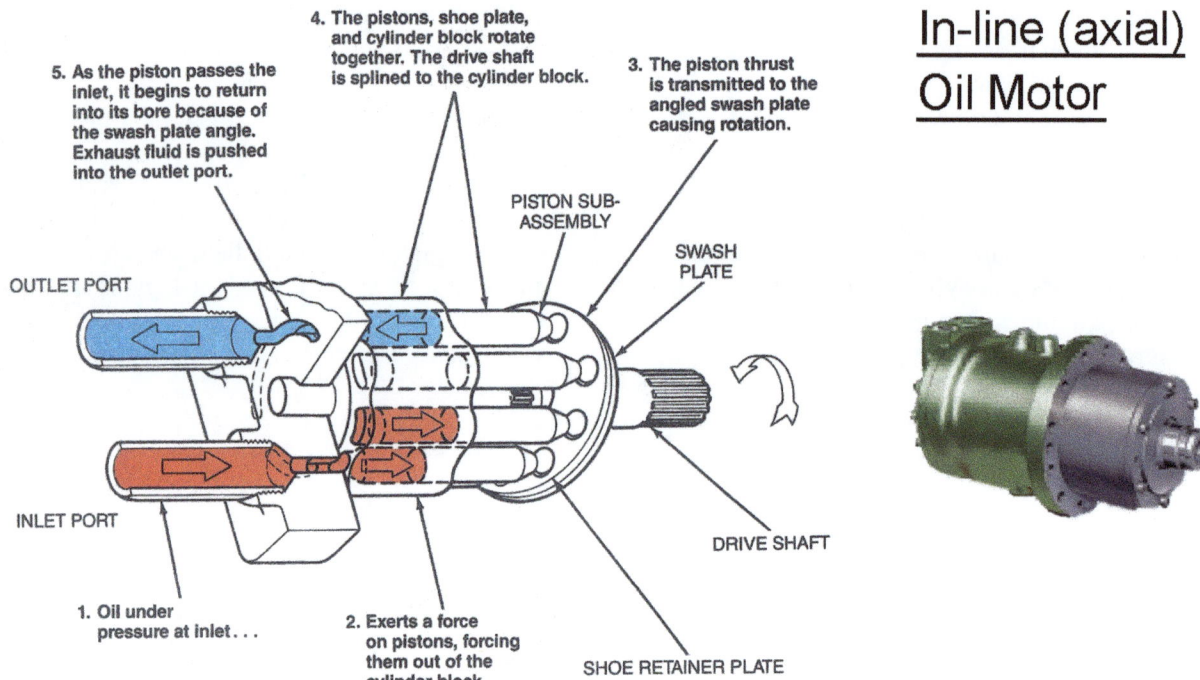

Rotary Actuator

A device for producing torque from hydraulic pressure and used to rotate the machine table. Inside the unit are either one or two vanes attached to a splined shaft which is connected to the rotation arm / machine rotary table. Hydraulic oil pressure is applied to one side of the vane(s) to produce the rotate action. Meter-out flow control valves are used to control the actuator rotation speed. The design of vane type rotary actuators does not allow an in-built cushion to slow down the table near the end of rotation and so an external cushion valve is used to prevent any damage at the stopping point. External stops are used to provide precisely 90° or 120° rotation (depending upon whether it is a four or three station machine).

Single-vane type rotary actuator

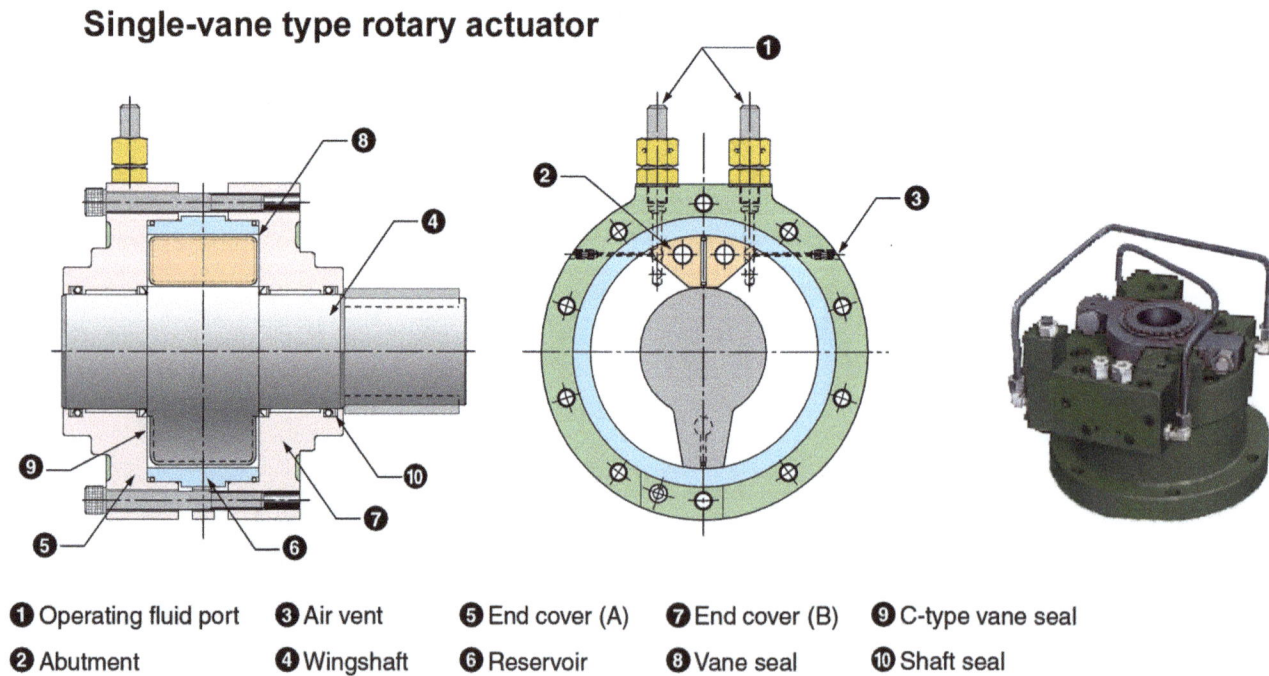

❶ Operating fluid port　❸ Air vent　❺ End cover (A)　❼ End cover (B)　❾ C-type vane seal
❷ Abutment　❹ Wingshaft　❻ Reservoir　❽ Vane seal　❿ Shaft seal

Hydraulic Cylinder

A device that converts the fluid flow produced by the hydraulic pump into a linear output so that it can perform useful work. Below is a cross section of a typical double-acting hydraulic cylinder. Fluid routed to and from the cylinder through ports in each end will cause the piston and the chrome plated steel piston rod assembly to move, giving powered motion both when extending and when retracting. Such cylinders are classed as differential cylinders because unequal areas are exposed to pressure during the extend and the retract movements. The difference is due to the cross sectional area of the rod which subtracts from the area under pressure during retraction. As a result extension is slower than retraction because more fluid is required to fill the swept volume of the piston, but greater force is available because the pressure operates across the full piston area.

When retracting, the same flow from the pump will result in faster movement of the cylinder because the swept volume is less. In the same way and despite having the same system pressure, the maximum force exerted by the cylinder will be less because of the smaller area under pressure.

The piston speed may be adjusted by controlling the flow of fluid either entering or leaving the cylinder. A *meter in* circuit is used where the force is needed to lift a load. If the load can run away from the actuator, a *meter-out* circuit must be used.

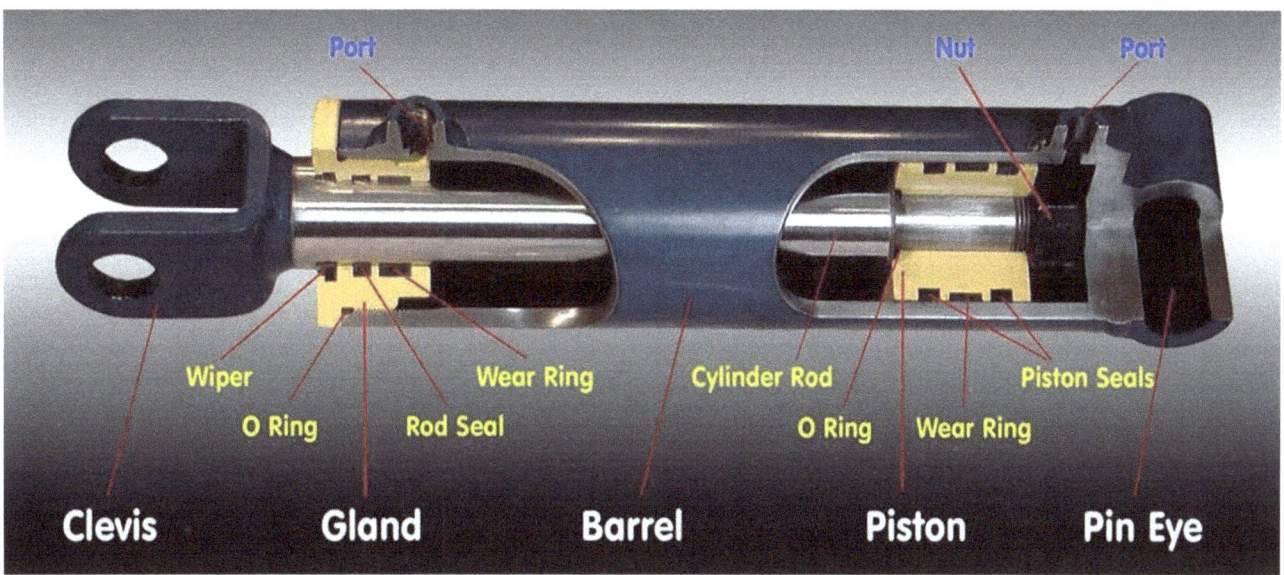

Maintenance

Many checks are simple and require no special tools or instruments. Pipe runs and hoses should be inspected for damage and to ensure all supports are intact and secure with no leaks. Connections subject to vibration should be examined for tightness and strain. Cylinder rods should be examined for score marks which can indicate dust ingress.

Regular checks on oil condition are of the utmost importance and any change in oil level should be investigated without delay. The oil temperature is continuously monitored and an alarm will sound if it becomes too high. Apart from the obvious heat exchanger faults (i.e. no water flow), other possible causes for excessive temperatures are incorrect operation of relief valves (i.e. the pump on load continuously), internal leakage, or too high a fluid viscosity.

System pressure should be recorded and checked against design values. Deviations can indicate maladjustment or potential faults. Too high a pressure setting wastes energy and shortens operational life. Too low a pressure setting may cause relief valves to operate at pressures below that needed by actuators, leading to no movement. Pressure deviation can also indicate developing faults outside the system. For example the fouling of a component moved by an actuator may cause a perceptible rise in pressure before failure occurs.

Contaminants, the natural enemy of hydraulic systems, cause more than 70% of all failures. If not controlled, particles too small to be seen can reduce hydraulic system efficiency (and system efficiencies can be reduced by as much as 20% before it is recognized that something is wrong).

Contamination affects hydraulic systems in many ways:

- Corrosion of hydraulic systems from acids that form due to fluid breakdown and mixing of incompatible fluids in the system.

- Increased internal leakage which lowers the efficiency of pumps, motors, and cylinders. It decreases the ability of valves to control flow and pressure accurately. It also wastes energy and generates excess heat.

- Sticking of parts due to sludge or silting (a sediment of fine particles).

- Seizure of parts or components caused by large amounts of contaminants getting stuck in the clearances.

There are several major sources for system contamination: (1) contamination built-in at the point of manufacture, (2) hydraulic fluid contamination, (3) environmental contamination, (4) component wear contamination, and (5) contamination introduced during servicing:

1) Built-in contamination, or primary contamination, is caused during the manufacture, assembly or testing of the hydraulic components. Metal fittings, small burrs, pieces of PTFE tape (or other sealing compounds), sand and other contaminants are routinely found during the initial clean up filtration of newly manufactured systems. These can be the most damaging particles to your system. Filtering them out immediately using a fine filter (finer than needed for normal operating life) will prevent early catastrophic system failure, and continuing system leakage problems.
2) Assume all hydraulic fluid is contaminated. Even "new" hydraulic fluid is contaminated to a level that is unacceptable for most hydraulic systems. Always filter new hydraulic fluid when filling a system, even when just "topping-off", otherwise contaminants will be introduced into the system with the new fluid. Check the hydraulic fluid to be sure it meets the ISO 4406 code[5] for the system in which it will operate.
3) Ingressed or external contamination comes from the environment. Dirt can enter the hydraulic system fluid supply through rod seals, breather caps, and worn cylinder rods.
4) Normal system operation results in wear of internal rod ends, valve spools, pump vanes, and even rubber hoses; generating minute particles contaminating the hydraulic system.
5) When systems are checked or disassembled for inspection or repair, they become vulnerable to contamination by dust and dirt adhering to filler caps, breathers, funnels, transfer pumps, and also the replacement parts themselves. Therefore great care must be taken during all repairs to keep everything clean:
    - clean exterior surfaces of dust, dirt, oil, etc. before removing covers
    - make sure new parts are clean
    - use a vigorous wash when cleaning parts
    - keep parts protected prior to assembly
    - protect system openings: use covers, tape, plastic wrap, etc.
    - clean transfer containers, funnels, nozzles, etc.
    - use covers to protect drums from becoming dirty or wet around their lids
    - filter fluids before filling the tank, don't remove filler screens
    - clean filters for all hydraulic fluid handling
    - use fine "clean-out" filters to clean the system after assembly

> **ACTION POINT**
>
> **Consider fitting *additional* oil filters in the hydraulic system.**
>
> **An external bypass filter will:**
> - **Reduce oil contamination**
> - **Increase the "life" of the oil**
> - **Reduce wear and tear on components**
> - **Improve system energy efficiency**
> - **Reduce maintenance costs**
> - **Reduce losses due to breakdowns**

Grease must be applied to provide a low-friction film between moving parts. A *food-grade* (physiologically safe and neutral in terms of odour and taste) grease should be selected due to the possibility of residue entering the stretch-blown container. The internationally recognised "H1" classification refers to lubricants which can be used for friction points with "incidental, technically unavoidable contact with the food [container]". Food-grade lubricating greases are typically derived from liquid paraffin combined with various performance-enhancing additives designed to counteract wear, prevent corrosion, reduce friction, improve adhesion of the grease, etc.

To ensure a satisfactory service life when selecting a grease, choose one whose upper operating temperature limit is considerably higher than the maximum temperature reached during normal service, since the "life" of a grease reduces significantly as its temperature rises (by an order of magnitude for each 50°C rise in temperature).

Sumico H1 Certified White Alcom Grease No 2

http://www.sumico.co.jp/e/pro_ind/use_food_e.html

Klübersynth UH1 64-62

http://www.klueber.com/en/product-detail/id/384/

---

[5] ISO 4406: 1999 Hydraulic fluid power—Method for coding the level of contamination by solid particles

6

Pneumatics

Pneumatic systems use the force of flowing gases to transmit power to where it is needed to produce mechanical motion. One-stage machines have two independent compressed air systems: lower pressure operation air to actuate the pneumatic cylinders, and the higher pressure blow air used to form the container (covered in detail in *Chapter 2—Injection stretch-blow moulding* and *Chapter 8—Plant Engineering*).

Pneumatic systems, unlike hydraulic systems that contain liquid, use compressed gas. Liquids are incompressible and can be pressurised almost instantaneously, whereas compressing a gas takes time. Since flow, and therefore pressure buildup in a hydraulic system can be quickly and easily controlled by means of unloading valves, hydraulic oil is stored in a reservoir at atmospheric pressure and pressurised only when needed. In contrast the slow response of an air compressor doesn't permit such an approach, meaning air must be compressed in advance and stored ready to use in a reservoir.

Pneumatic systems comprise assorted valves, actuators, and ancillary components connected by metal or plastic tubing. Pneumatic equipment is straightforward to operate, is generally reliable, and requires little maintenance (other than keeping it, and the compressed air supply, clean and dry). However pneumatic systems cannot drive loads that are too heavy, have poor speed control, relatively low accuracy, and exhaust air can be very noisy if it's not correctly vented.

In a machine, every item is located to properly perform its function. In the Manual however, pneumatic systems are represented schematically: graphical symbols depicting component function are linked by lines corresponding to functional relationships. Schematics are intended to elucidate system logic, and sometimes bear scant relation to actual layout. In machine manuals reference numbers for pneumatic valves and cylinders are prefixed with the letters *AV* (and *YAV* for solenoid operated direction valves in the Aoki manuals).

Directional control valves have two or more positions, and so the symbol for a directional control valve is formed of one or more squares each representing a possible valve position. By convention, the inlet and exhaust are shown underneath the square, while the outlets are at the top. Arrows "↓↖" are used to indicate the direction of the air flow. If an external port is not connected to the internal parts, the symbol "⊤" is used. The symbols for the valve actuators (typically either a spring or a solenoid) are drawn on the outside of the squares.

It is important to remember that a pneumatic circuit diagram represents the circuit in static form, assuming there is no supply pressure.

Chapter 6—Pneumatics

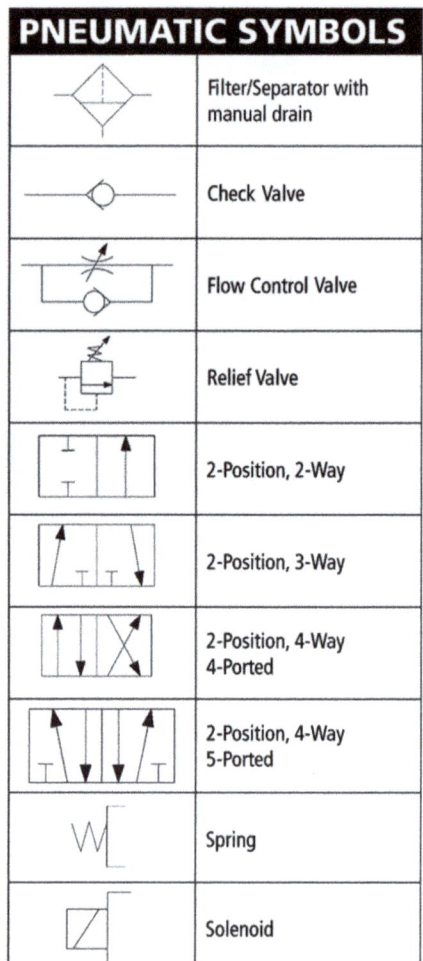

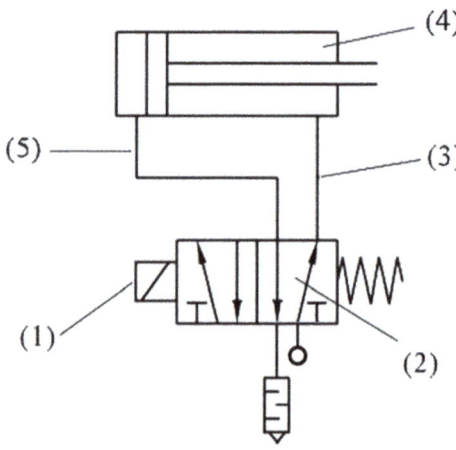

Operation air

Operation air entering the machine first goes through a residual pressure exhaust valve which automatically exhausts the operation air system should the emergency stop be pressed, and then passes through a Pressure Regulator and a Filter.

The pressure regulator reduces the factory supply pressure to the ~1 MPa required by the machine and maintains this pressure regardless of inlet pressure fluctuations. A pressure regulator is not intended to be used as a shutoff device. When the machine is not in use, the incoming supply should be turned off.

The Air Filter channels the air in a swirling pattern to fling out any liquid against the sides of the polycarbonate bowl. The bowl needs to be checked regularly and emptied when necessary (taking care to collect the emerging condensate).

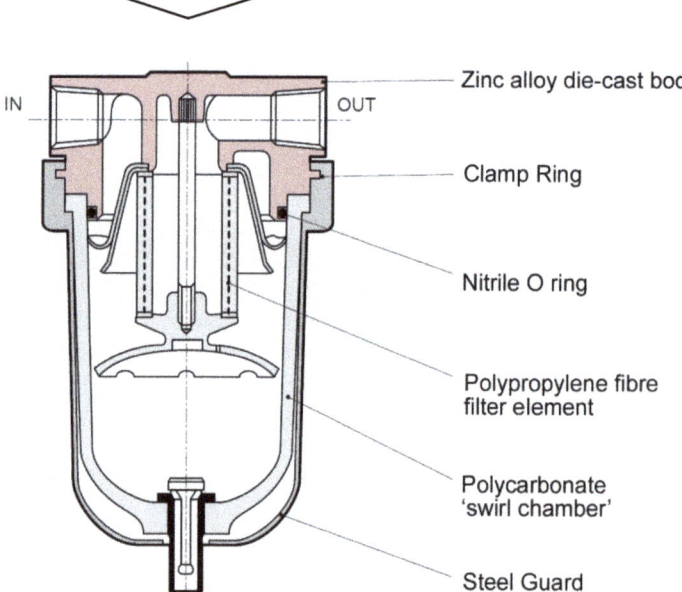

1. When the solenoid is not energised, the spring will restore the valve to its "home" position.
2. From the position of the spring, one can deduce which "square" is operating, (the other "square" does't come into play until the solenoid is energised).
3. Air pressure exists along this line because it is connected to the source of compressed air.
4. As this cylinder cavity and piston rod are under the influence of pressure, the piston rod is in its retracted position.
5. The rear cylinder cavity and this line are connected to the exhaust, where air is released.

Double-acting pneumatic cylinders

The double-acting cylinder is the most common type in industrial applications. Pneumatic pressure can be applied to either port to give powered motion both when extending and when retracting. Pistons are made of cast iron with polyurethane nitrile rubber seals. Piston rods are hard chrome plated steel for wear and corrosion resistance. End caps are mild steel with threaded entries for ports. The cylinders themselves are either made from seamless drawn steel or extruded aluminium tube with a super-smooth inner wall so that they do not require lubrication. To remain effective though, these surfaces must maintain their mirror-like smoothness. Contamination can damage the smooth surface, which means that non-lubricated pneumatic systems demand consistently dry air and reliable filtration.

Chapter 6—Pneumatics

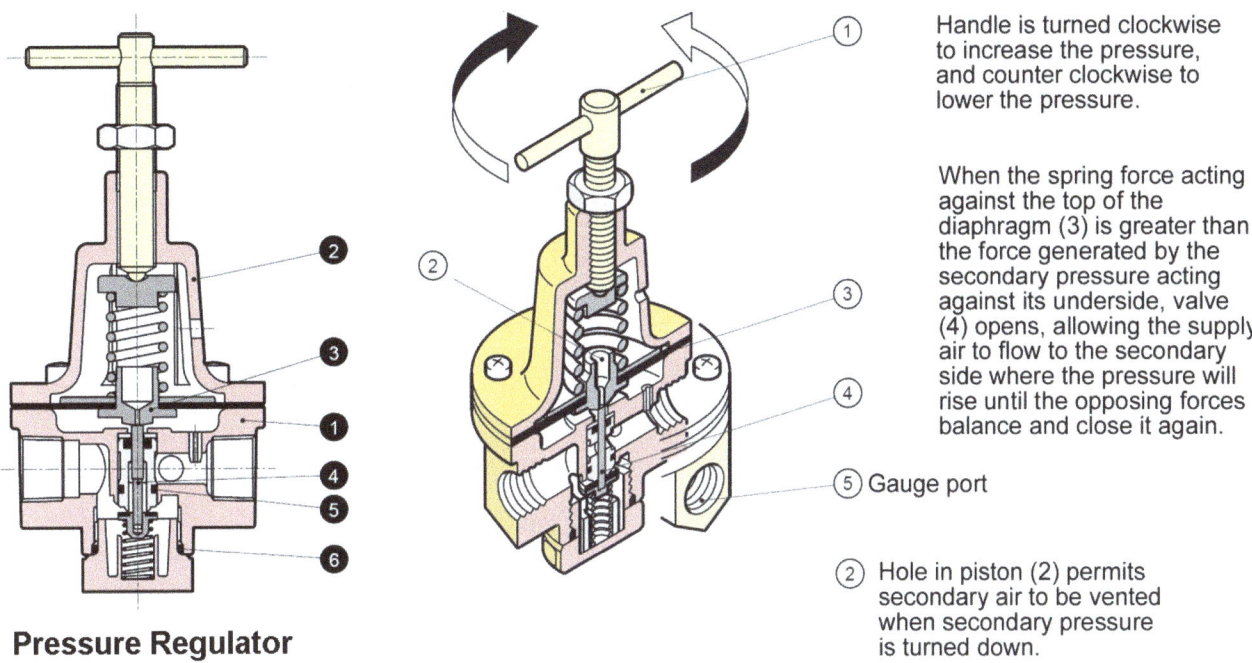

Pressure Regulator

Handle is turned clockwise to increase the pressure, and counter clockwise to lower the pressure.

When the spring force acting against the top of the diaphragm (3) is greater than the force generated by the secondary pressure acting against its underside, valve (4) opens, allowing the supply air to flow to the secondary side where the pressure will rise until the opposing forces balance and close it again.

(5) Gauge port

(2) Hole in piston (2) permits secondary air to be vented when secondary pressure is turned down.

No.	Description	Material
1	Body	Zinc die casting
2	Guard	Zinc die casting
3	Diaphragm assembly	Brass, steel, nitrile rubber
4	Valve assembly	Stainless steel, brass, hydrogen nitrile rubber
5	O ring	Nitrile rubber
6	O ring	Nitrile rubber

NB: A Pressure Regulator valve should be set slightly below the supply pressure so that it can maintain the operation air at a constant pressure, independent of any fluctuation in the supply!

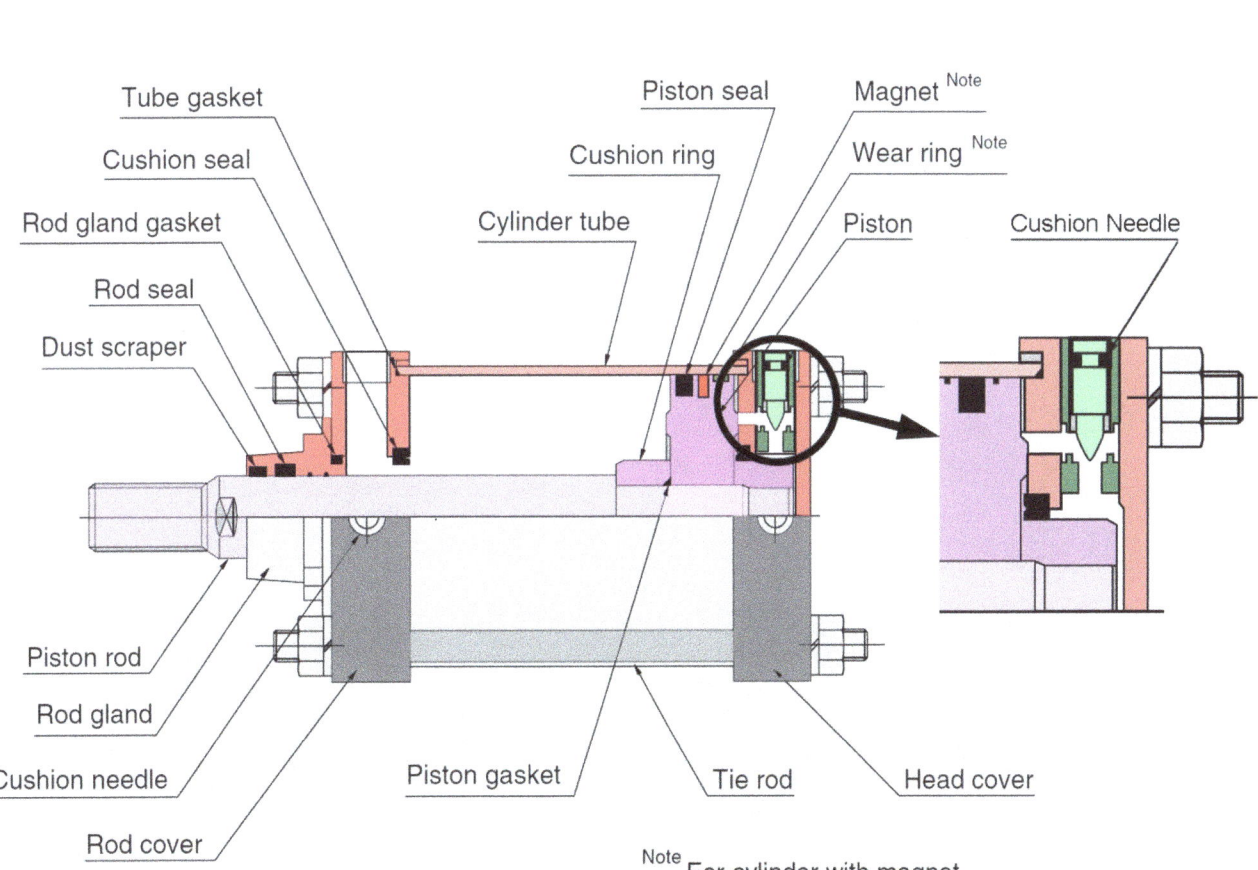

Note For cylinder with magnet

Cushioning can be provided to decelerate the cylinder load as it approaches the end of stroke and prevent the piston from banging into the end cap which could damage it and reduce the service life of the actuator. Cushions are simply small diameter pistons that enter a shallow receptacle machined into the end caps. When the cushion piston enters the cushion sleeve, the flow of air leaving the sleeve can be restricted to slow the speed of the actuator.

The machine's control system needs to 'know' the position of the piston (or the movable mechanism) and so pneumatic cylinders are either fitted with magnetically operated proximity switches mounted on the outside of the (non-magnetic) cylinder, which are operated by the magnetic field of a permanent magnet integrated into the piston, or separate proximity sensors are built into the framework of the movable mechanism itself.

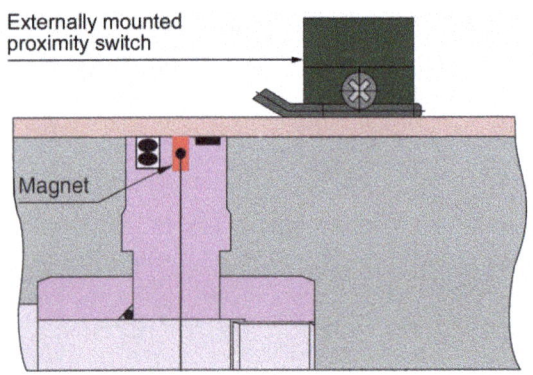

Solenoid Operated Directional Valves

Directional valves control the flow of air by opening and closing and air ports. They are classified by the number of ports, the number of switching positions, the normal position of the valve, and its method of operation. Common types of directional control valves include 2/2, 3/2, 5/2, etc. The first number represents the number of ports, and the second is the number of "positions". The symbol in the drawing on *Page 64* represents a directional control valve with five ports and two positions.

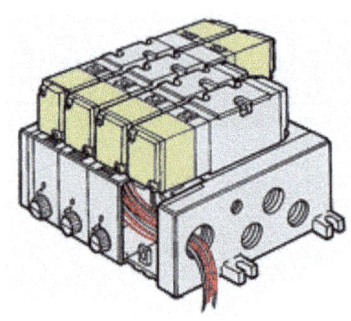

From an efficiency perspective, it might seem that the ideal location for a directional valve would be on the actuator itself as the volume of air in the pipe between the valve and cylinder is minimized, but "manifolding" pneumatic solenoid valves together as shown above has a number of advantages:

- Manifolded valves lead to cleaner, more integrated control packages, with just one or two supply and exhaust points for several valves.
- Electrical inputs are close together in one place.
- A valve manifold typically allows easier valve mounting and valve replacement than do body-ported valves.

Most solenoid operated directional valves that are used in pneumatic systems are *spool* valves because they offer a good balance of reliability and durability versus cost. A *spool* is an internal element of a valve that opens or closes flow paths by sliding back and forth in the bore of the valve.

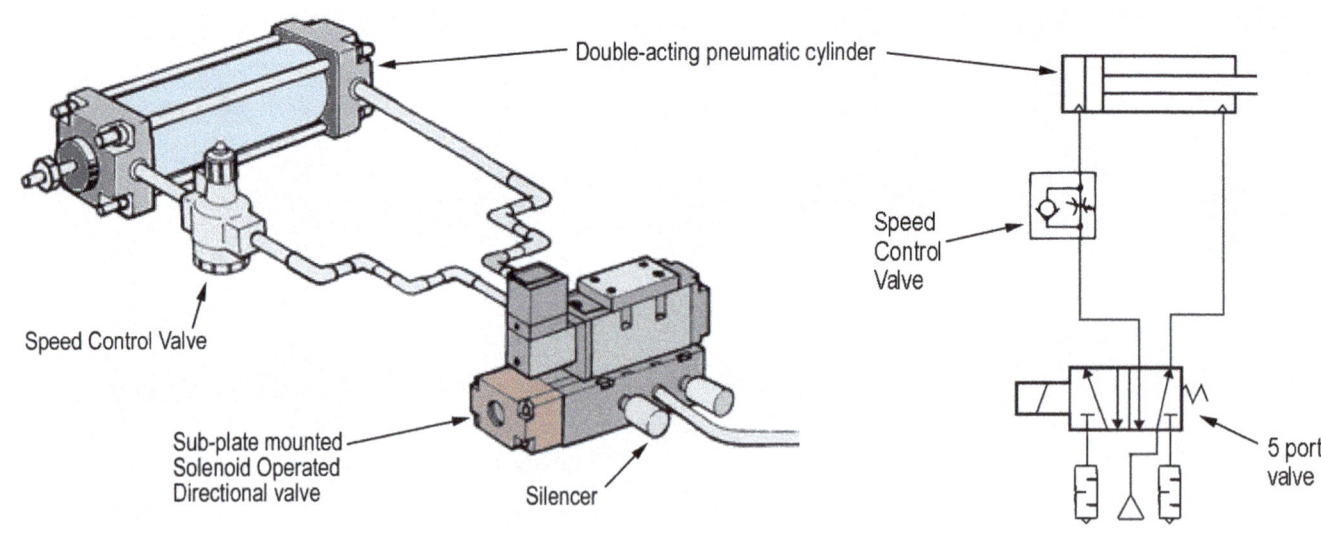

Two–position, single solenoid directional valve

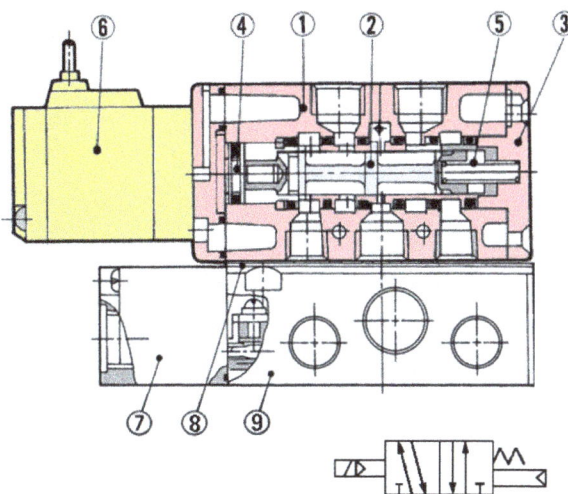

No.	Description	Material
1	Body	Aluminium die-cast
2	Spool/ Sleeve	Stainless steel
3	End plate	Resin
4	Piston	Resin
5	Return Spring	Stainless Steel
6	Solenoid Assembly	
7	Junction Cover	Resin
8	Gasket	Nitrile butadiene rubber
9	Sub-plate	Aluminium die-cast

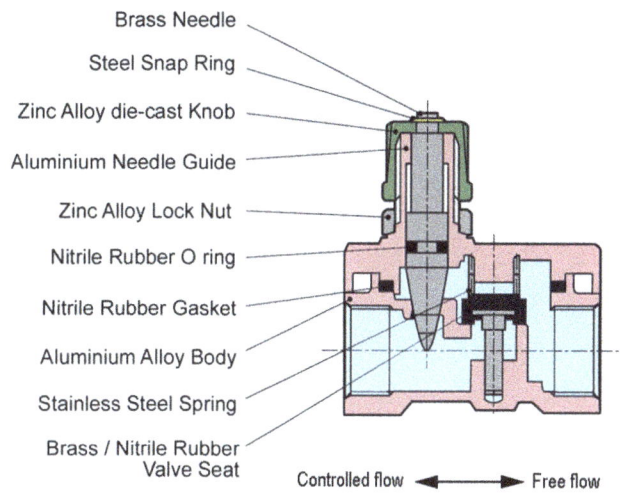

Speed Control Valve

A Speed Control Valve allows free flow in one direction, and adjustable or restricted flow in the other direction. Typically they are used to control the speed of a cylinder in one direction, by restricting the flow of exhaust air out of the cylinder. Examples include the *Blow Core Down*, and *Ejector* cylinders.

In one direction the air flow is controlled by the needle valve, in the other the poppet opens to permit free flow. Note that the compressibility of air — advantageous where smooth operation is concerned — makes precise speed control rather more difficult for pneumatic than hydraulic systems.

Quick Exhaust Valve

The Quick Exhaust Valve allows the air to escape from the rod end of the pneumatic stretch cylinder more rapidly , thus enabling a faster *stretch rod down* speed.

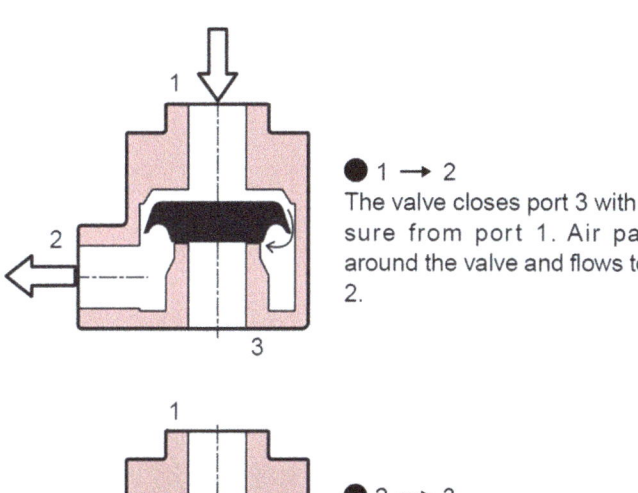

● 1 → 2
The valve closes port 3 with pressure from port 1. Air passes around the valve and flows to port 2.

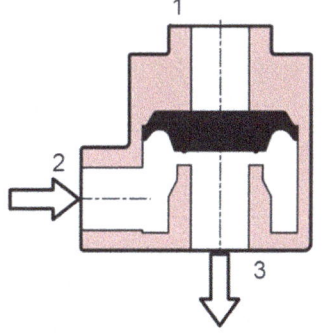

● 2 → 3
When port 1 pressure drops, the valve closes port 1, opens port 3, and exhausts port 2 air.

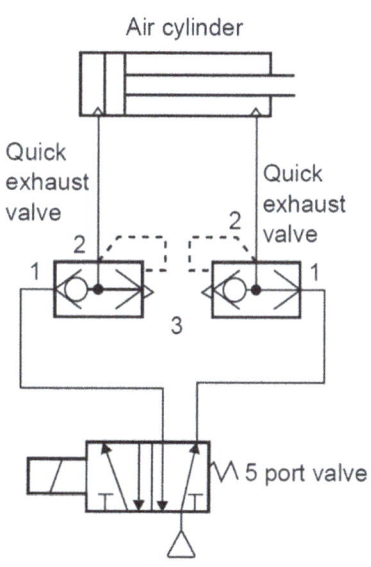

7

Electrics

The word *electricity* derives from the Greek word for amber — the fossilised resin which when rubbed produces a static charge that makes lightweight objects cling to it. With no obvious practical application, it was not until the invention of the battery at the beginning of the nineteenth century provided a more reliable source of electricity that the link between electricity and magnetism was recognised. Then once it had been discovered that an electrical current was induced in a wire moved through a magnetic field, it wasn't long before electricity was being generated and distributed.

Typically mains electricity is distributed via three conducting wires and a neutral/ earth connection. Each conductor carries alternating current (AC) with the same frequency (50Hz)[6] and voltage (400V)[6] but with a phase difference of one third of a cycle. This means that providing electrical loads are evenly distributed between the three conductors, no net current flows in the neutral. This system can provide both 400V three-phase and 230V single-phase power: loads connected between phases are fed at 400V, while those connected between one phase and neutral receive 230V (making it suitable for supplying both industrial and domestic consumers). Ideally each of the three conductors will supply approximately the same current load. Consider an

---

[6] Europe: $400/\sqrt{3} = 230V$

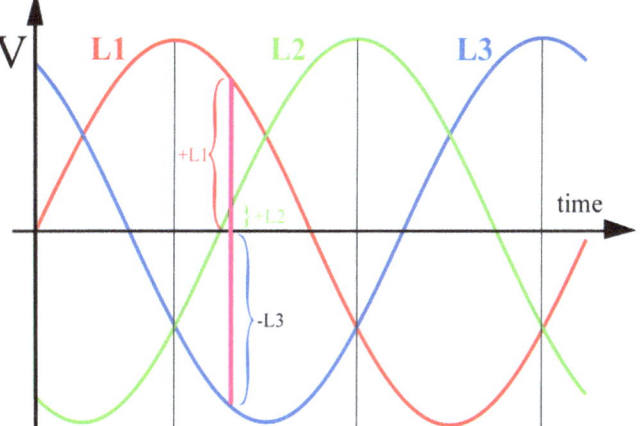

Where the loads are equal on each of the three phases, no net current flows in the neutral: **L1** + **L2** + **(-L3)** = zero

ISBM machine: the induction motor requires a 400V three-phase supply while the injection cylinder heater bands require a 230V single-phase supply. (Note that for a given size of cable, more power can be supplied by a three-phase supply than is possible with a single-phase supply). Fortuitously three-phase motors are load-balanced by design (since the resistance of each phase winding is the same), but to ensure the heater band loads are balanced, their electrical connections must be allocated between neutral and all three phases.

Transformer

A major advantage of AC is the ease with which an alternating voltage may be stepped down (reduced), or stepped up (increased) by a transformer. This enables mains transmission systems to function at very high voltages (higher voltages mean lower currents, making possible the use of smaller sized cabling and switchgear, and reduced power losses). These high voltages are then stepped down at local electricity sub-stations for distribution to consumers.

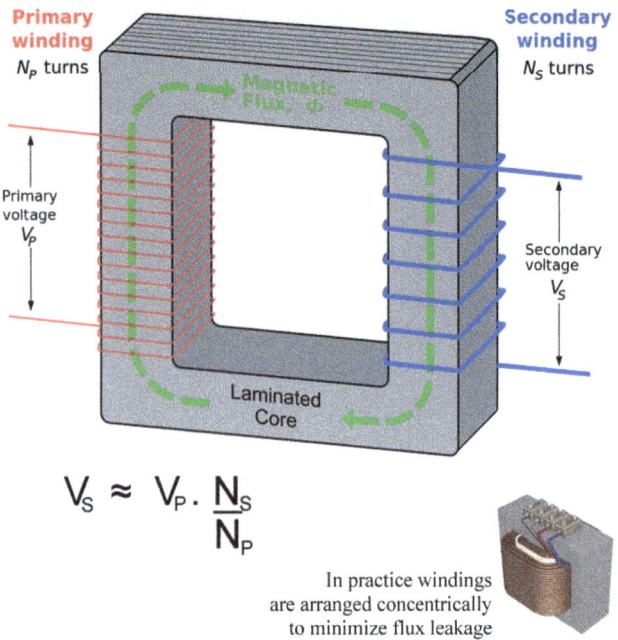

$$V_S \approx V_P \cdot \frac{N_S}{N_P}$$

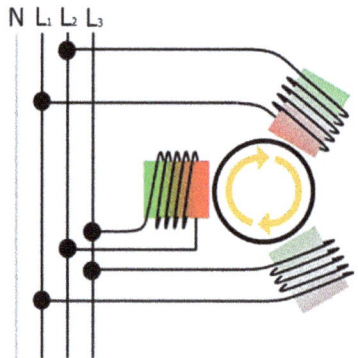

In practice windings are arranged concentrically to minimize flux leakage

A varying current in the transformer's primary winding creates a varying magnetic flux in the laminated steel core, which in turn induces a varying voltage in the secondary winding. The magnitude of this induced voltage is a function of the number of turns in the primary and the secondary windings.

On ISBM machines several transformers coupled with rectified power supplies provide power appropriate to each component, such as 24V DC for the PLC and most of the control elements (the one exception being the blowing air system where 100V AC solenoids are used to provide higher speed operation of the oversize pneumatic valves).

Induction motor

When a three phase supply is connected to the windings equally distributed around the stationary part or *stator* of an induction motor, the alternating current produces a magnetic flux rotating at the same speed as the supply frequency. Its direction of rotation can be reversed by exchanging any two phases (e.g. L_1 to L_2 and L_2 to L_1). The effect of this rotating flux is to induce opposing currents to flow in the conductors of the *rotor*, which in turn produce an opposing magnetic flux, the end result being the application of a torque to the rotor that makes it rotate in the same direction as the magnetic field.

However until the rotor has built up sufficient speed for the two opposing fluxes to "balance", the stator could draw such a high current that its windings would overheat (as well as causing a voltage drop on the supply line that could be harmful

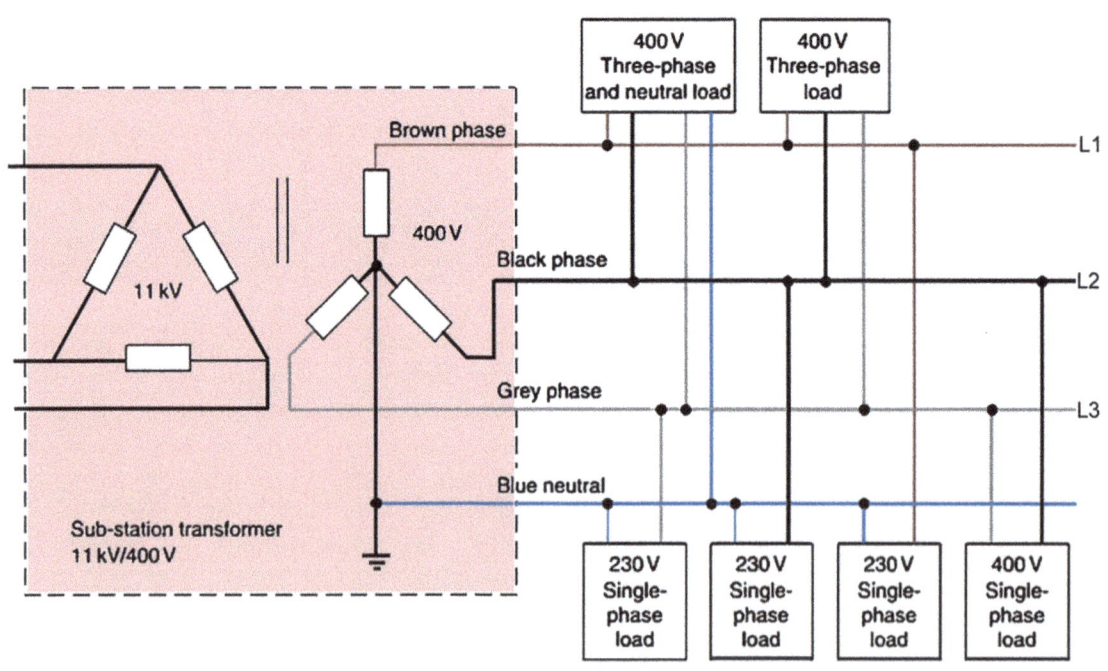

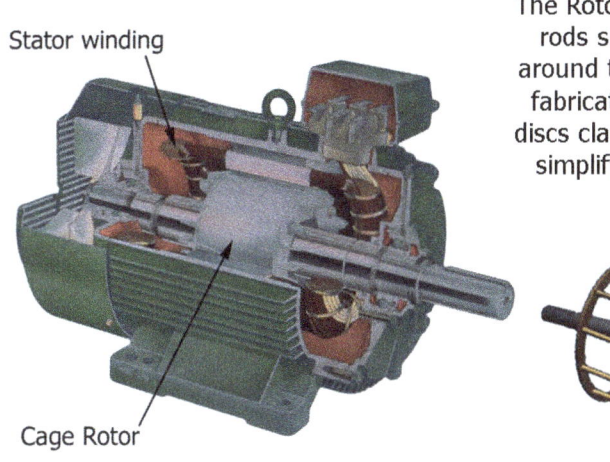
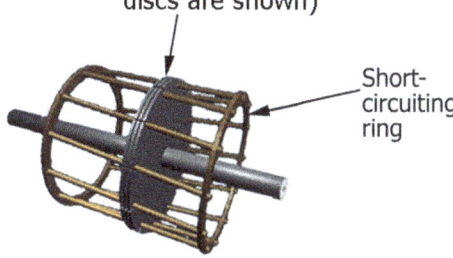

Stator winding

Cage Rotor

The Rotor comprises copper or aluminium rods slotted into holes equally spaced around the circumference of an iron core fabricated from multiple insulated steel discs clamped together (for clarity and to simplify the drawing, only three of the discs are shown)

Short-circuiting ring

to nearby equipment). The *motor starter* is designed to prevent such heavy starting currents. When the motor is first switched on the stator windings are initially connected in *star*, and then a few seconds later once the rotor is spinning, the connection is automatically switched to a *delta* configuration. Connecting the stator windings in *star* reduces the phase voltage to about 58% of the running voltage (and also reduces the motor's torque). Then once the motor is up to speed a double throw switch makes the changeover from *star* to *delta*, thereby achieving the twin aims of a minimum starting current with maximum running torque.

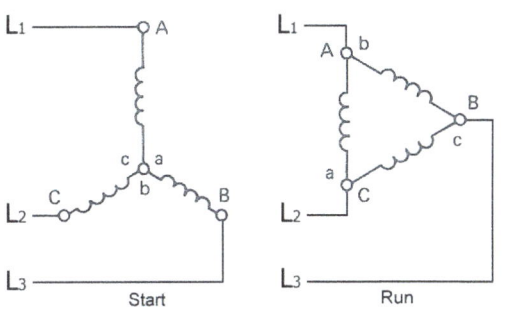

Motor winding connections
Star - Delta Starter

When an induction motor is switched on, the rotor speed steadily increases until it approaches that of the rotating magnetic flux — although it never quite matches it because if it did so, there would be no induced emf. For this reason induction motors are also referred to as *asynchronous motors*.

On ISBM machines the induction motor is protected by a *thermal overload relay* which causes the starter contacts to open should the current flow exceed the pre-set value. Assuming that there is no continuing problem, once the bi-metallic element inside has cooled the reset button can be depressed and then the motor restarted in the normal manner.

Asynchronous motors operate at a constant speed chosen by the machine designer to satisfy the maximum hydraulic demand. This is inefficient as demand for oil varies during the cycle and so aftermarket suppliers now offer *variable-frequency drive* (VFD) controllers which control the frequency of the electrical power supplied to the motor and thus its rotational speed. VFD controllers are solid state electronic power conversion devices that first convert the AC input power to DC "intermediate" power using a rectifier. The DC intermediate power is then converted by an inverter switching circuit back into the quasi-sinusoidal AC power supplied to the motor. A control signal from the one-stage machine to the VFD is used to trigger one of two preset motor speeds — a higher speed during the maximum hydraulic demand, i.e. when the moulds are opening and closing, and a lower speed during periods of reduced demand. The use of variable-frequency drives eliminates the need for a Star-Delta starter since the inverter circuitry controls the electrical current delivered to the motor.

Programmable Logic Controller

Until the early 1980s machine control, sequencing, and safety interlock logic were implemented by means of multiple electro-magnetic relays, cam timers, and dedicated closed-loop controllers. The major disadvantage of this approach was that mechanical components are prone to wear, leading to untimely failure, while any sequence changes that became necessary involved wholesale rewiring. Following the development of the integrated circuit, programmable logic controllers (PLC) were introduced. A PLC is a micro computer running a program which controls the operational sequence of the machine, i.e. when and in what order movements occur. Unlike personal computers, the program is stored in non-volatile memory.

All but the smallest PLC are modular in construction and mounted in a rack into which additional modules with dedicated functions such as a power supply, relay units for input (signals from switches and sensors) and output (sending on/off signals to solenoids), and also temperature controllers may be plugged. A high speed serial link enables the modules to communicate, and also allows additional racks to be situated remotely from the processor in a more convenient location and using less complex wiring.

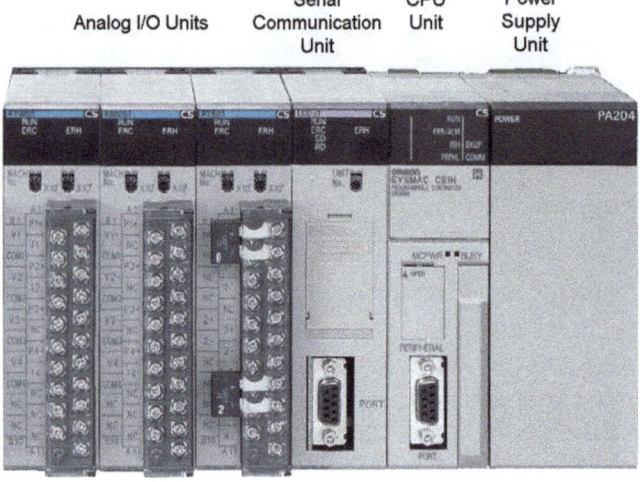

In the first PLC controlled machines, the PLC was directly connected to each signal generator (limit switch, temperature controller, transducer), and each actuator (solenoid) via dedicated 24V cables, while devices such as visual displays (when fitted) were separate from the PLC. Nowadays this is changing with controllers communicating with each signal generator and control element over a network, yet many existing production machines continue to be controlled by a dedicated stand-alone, directly connected PLC.

PLC are typically programmed in *ladder logic*, a form of notation which resembles a schematic diagram of relay logic (originally chosen because it would appear familiar to existing technicians, and still used today because PLC logic runs in a predictable and repeating sequence, and the schematic format enables the programmer to pick out any issues with the logic and sequence timing more easily than would be possible in other formats). However programs are actually written, and execution simulated using software typically running in a Windows environment.

A ladder diagram consists of left and right *bus bars*, connecting lines, input bits, output bits, and special instructions. A program consists of one or more program *rungs*. Instructions are executed in order starting with the top rung, and reading from left to right, in a single cycle.

In operation the PLC continually scans each rung of the program. First *examining* each input to determine if it is on or off, before *executing* the program one instruction at a time. For example, a program might state that if Input 001 is ON, then Output 006 should be energised. In the third step the PLC finally *updates* the status of the outputs, in this case (assuming Input 001 had been ON) actually energising Output 006. After the third step the PLC restarts at the beginning and repeats the steps in an iterative process.

The ISBM process involves numerous machine movements which must all occur at the proper time. To control these movements *timers* are simulated by the CPU. Timers come in several varieties, with the most common being the *on-delay* (i.e. countdown) type. *Off-delay* timers are used to control decompression times and barrel shut-off nozzle closing time (only when an off-delay timer is *de-energised* does it start counting, and the relevant output is triggered when it times out).

An area where timing is critical is the delay between the stretch rod being triggered and the start of the primary (P1) blow, because this time gap is largely responsible for determining the material distribution up and down the bottle. Since a small change can mean the difference between an acceptable container and scrap, this *on-delay* timer can be incremented in steps of one hundredth of a second.

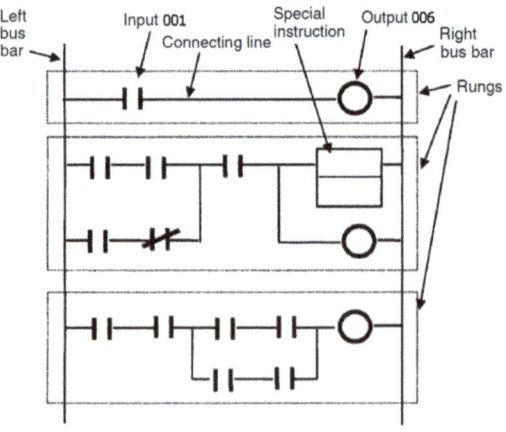

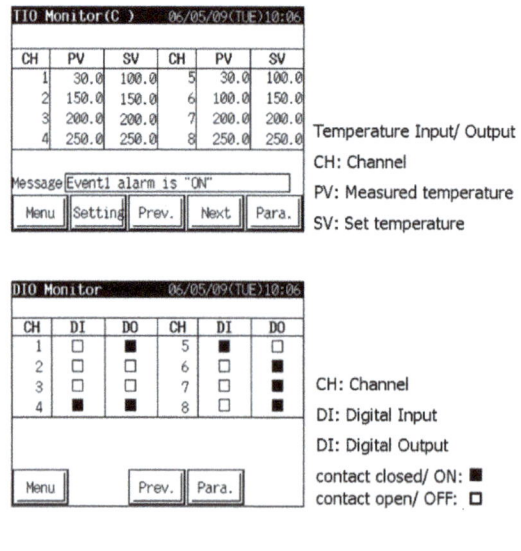

Typical operation panel providing temperature monitoring and setting (top), and monitoring of digital inputs and outputs (bottom).

Heater Bands

The heating of the polymer is carried out not just through the mechanical action of the screw, but also by *heater bands* which are electrical resistance heaters secured around the barrel, nozzles, and hot runner manifold. Heater bands are connected between one phase of the three phase mains supply and neutral in order to provide them with a 220-240V supply.

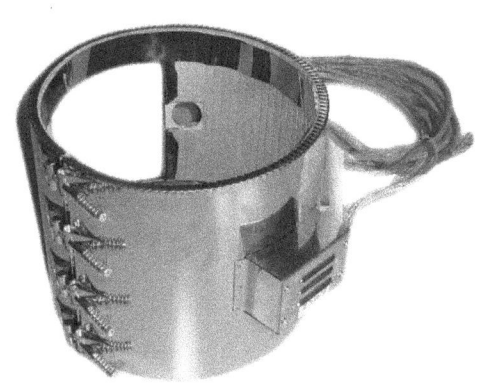

Since heater bands transfer heat through contact, they have to be a tight fit on the barrel (which itself must be clean and smooth to ensure efficient heat transfer). Contact may be further improved by smearing the mating surface with a copper-based paste, but as most heater bands have sheet metal cases, when first fitted they tend not to make good contact across their full area. What is more, the initial clamping force tends to relax as the band gets hot and expands, and so recently fitted heater bands should be re-tightened after the set temperature has been reached.

Important things to know about mineral insulated heater bands include:

- Heaters that are new, or have been stored at room temperature for a few days, typically have an insulation resistance measured between the leads and the sheath of only 10 - 50 kΩ (thousands of Ohms). Low insulation resistance values like this usually indicate the presence of moisture that has been absorbed by the mineral insulation during shipment and handling. Until the absorbed moisture is evaporated, mineral insulated bands will sometimes *trip* a controller equipped with fault detector. Setting a mineral insulated heater to 100°C for one hour will evaporate off any moisture. The insulation resistance should then be in the MΩ range (as seen in mica insulation which does not absorb moisture).

- Always replace a faulty Heater Band with one of the same size and power (Wattage) because the power of a heater is matched to the load requirement to limit constant on-off cycling.

- A significant amount of heat escapes into the atmosphere, and so fitting an insulating blanket to the barrel can be cost-effective.

Electro-mechanical relay

Relays

Relays are remotely controlled electro-mechanical switches that enable a low voltage/ low current circuit to switch a second high voltage/ high current circuit that is electrically isolated from the first.

The relay's terminals are typically labelled:

- COM = Common

- NO = Normally-open means the circuit is disconnected when the relay is inactive. (The contacts only connect the NO terminal to the COM terminal when the relay is energised).

- NC = Normally-closed contacts mean the circuit is connected all the time that the relay is *not* energised, only being disconnected from the COM terminal when the relay is energised. (Note that most electro-mechanical relays have more than one pair of NO and NC terminals).

A *solid state* relay (SSR) is a solid-state device providing a similar function to an electro-mechanical relay but has no moving components for greater long-term reliability.

Solid State Relay

Thermocouples

Thermocouples are used to measure the temperatures of the barrel, nozzle and hot runner system. Thermocouples consist of two dissimilar metal wires welded together at the end where the temperature is to be measured (the hot junction), while the free ends are connected to the terminals of a Temperature Controller (the cold junction) forming a circuit. When the temperature at the hot junction differs from that at the cold junction (i.e. so that there is a temperature gradient along both wires), a very small electric current[7] proportional to the temperature difference between the two junctions flows in the circuit.

The Temperature Controller interprets the current as a temperature which it displays digitally. To compensate for ambient temperature changes (which would otherwise cause inaccuracies in the temperature measurement), special thermocouple leads which alter their electrical resistance according to their temperature are used. Classic one-stage machines use only NiCr and NiAl wires, which are referred to as *Type K* thermocouples.

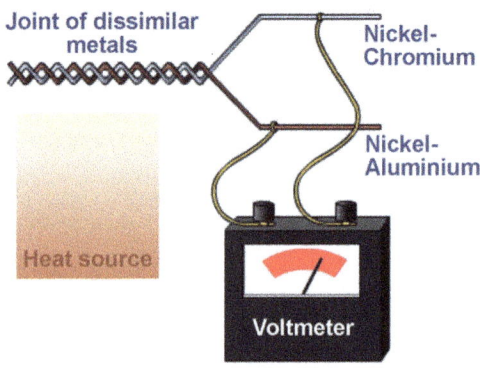

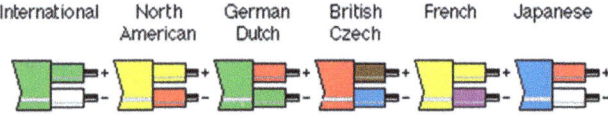

(ensure correct polarity when connecting thermocouple wires)

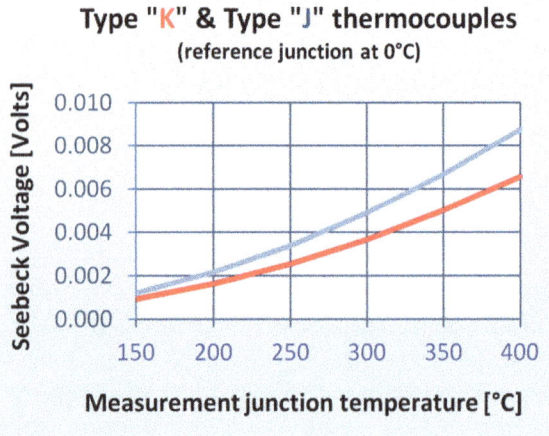

It is common to also find *Type J* thermocouples in plants which have injection moulding machines. They must not be confused with Type K because for a given temperature, Type J thermocouples generate a larger *Seebeck voltage* which means that if they were to be used by mistake, the temperature displayed would be hotter than it actually was.

Temperature Controllers

Temperature Controllers are closed-loop feedback controllers used to switch ON or OFF the solid state relays (SSR) which deliver power to the heater bands. A feedback controller is designed to generate an output that causes some corrective effort to be applied to a process so as to drive a measurable process variable towards a desired value known as the set point. Closed-loop feedback controllers determine their output by observing the error between the set point and the actual process variable measurement, while also taking into account how the process responds to the controller's corrective efforts. They are referred to as PID controllers owing to the three term closed-loop feedback control algorithm on which they operate. This utilizes proportional, integral and derivative terms, together with tuning constants reflecting the hysteresis of the system, to generate the controller's output. The proportional element is the size of the error now, the integral element is the cumulative effect of the error over recent time, and the derivative element is determined by how rapidly the error has come about. This means that if the current error is large, has been sustained for some time, or is changing rapidly, a PID controller will attempt to make a large correction by producing a large output. Conversely, if the process variable has matched the set point for some time, a PID controller will leave well enough alone.

Tuning a PID controller, in other words determining the appropriate constants to take into account the thermal characteristics of the system, is a skilled job, but fortunately modern controllers have an in-built auto-tuning capability. Nowadays, instead of having stand-alone controllers for each heater band, rack-mounted modular units are linked to a common display.

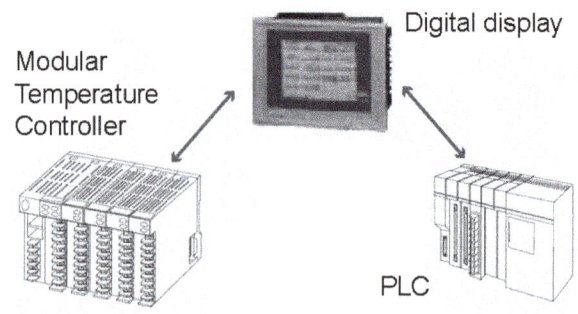

---

[7] Thomas Seebeck (1770 - 1831) discovered this phenomenon in 1821.

Voltage Regulators

Voltage Regulators are sometimes used to deliver an operator-set voltage to a heater band. The voltage may be increased to raise the temperature and reduced to lower it. Since there is no thermocouple measurement of the actual temperature of the heater band, the correct voltage setting, i.e. the one that gives the best outcome, is found by a process of trial and error. Voltage regulators can be used for the barrel and Hot Runner nozzles, the Sprue Bush and (when used) the heating pieces at the conditioning station of a four station machine.

Solenoids

Solenoids are electro-mechanical devices that convert the electrical signal from the PLC into linear mechanical motion in order to actuate the pneumatic and hydraulic directional valves.

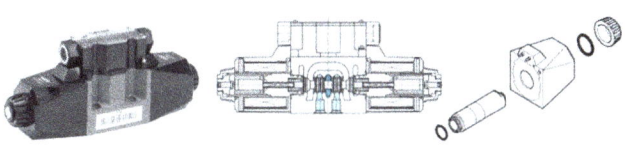

Each solenoid requires an electrical current that is proportional to the size of valve that it operates. In those cases where a solenoid requires a larger current or voltage than the PLC can supply, its energy is supplied indirectly, through a relay controlled by the PLC. Solenoids can operate on AC (usually 110 V AC) or DC (usually 24V DC). AC and DC solenoids have totally different operating characteristics. An AC solenoid has a very high inrush current producing a high initial force on the pilot spool. As the spool moves in, the inductance of the coil rises and current falls to a low holding current (and a low force on the pilot spool). Should the pilot spool jam, the current remains high, causing the circuit breaker to open (or the solenoid coil to burn out if the protection is inadequate). In contrast the current in a DC solenoid is determined by the coil resistance and does not change with pilot spool position. This means a DC solenoid delivers a lower force on the pilot spool, but will not burn out if the spool jams. Also the current in a DC solenoid tends to be higher, thus requiring larger diameter cables.

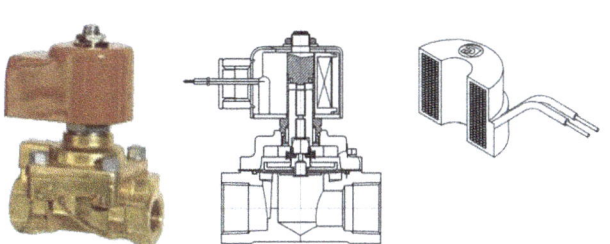

Solenoid operated directional valves are often fitted with a manual override for operating the valve without energizing the solenoid.

No Fuse Circuit Breakers

A no fuse circuit breaker is a device that automatically disconnects an electrical circuit in the event of excessive current flow. As its name suggests, it does not contain a fuse (which would need to be replaced if blown), but can be manually reset to resume normal operation.

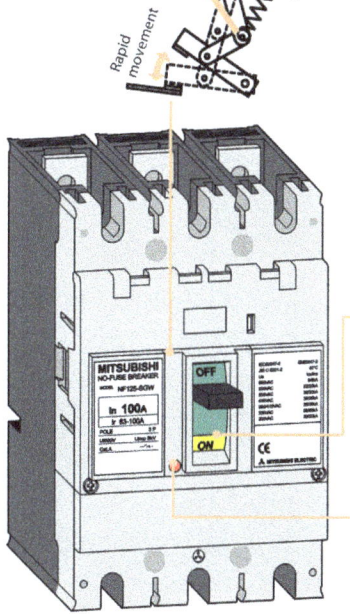

Switching mechanism
The contacts open and close rapidly, regardless of the moving speed of the handle, minimizing contact wear and ensuring safety.

Handle
- *Trip indication*
The automatically tripped condition is indicated by the handle in the center postion between ON and OFF.
- *Resetting*
Resetting after tripping is performed by first moving the handle OFF position to engage the mechanism, then returning the handle to ON to reclose the circuit.

Trip button (push to trip)
Enables manual tripping to confirm the operation of the manual resetting function.

Switches

Selector and *Push Button* switches are panel-mounted manually operated switches used to choose between options or to trigger machine actions. Note that rather than being direct acting, all manually operated control switches actually operate through the PLC. The *Emergency Stop* buttons installed on both the operator's side and the rear of the machine are self-latching panel-mounted or surface-mounted push button switches that once actuated, can only be released by twisting the head of the switch.

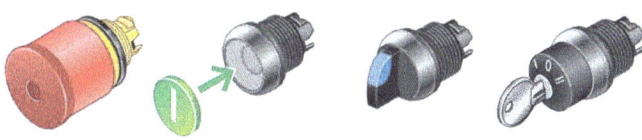

Control and Limit switches

Limit Switches sense the position of movable parts of the machine and may be either mechanically actuated devices such as the *Drop Out Confirmation*, or non-contact devices such as the proximity sensors which detect the position of the Upper and Lower Moulds.

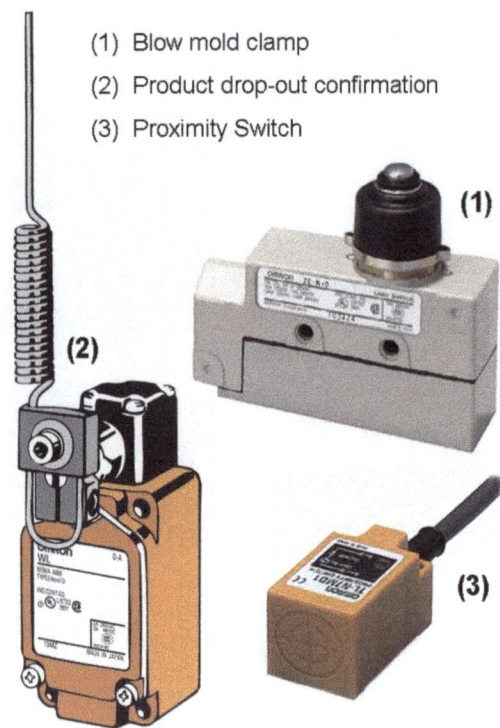

(1) Blow mold clamp
(2) Product drop-out confirmation
(3) Proximity Switch

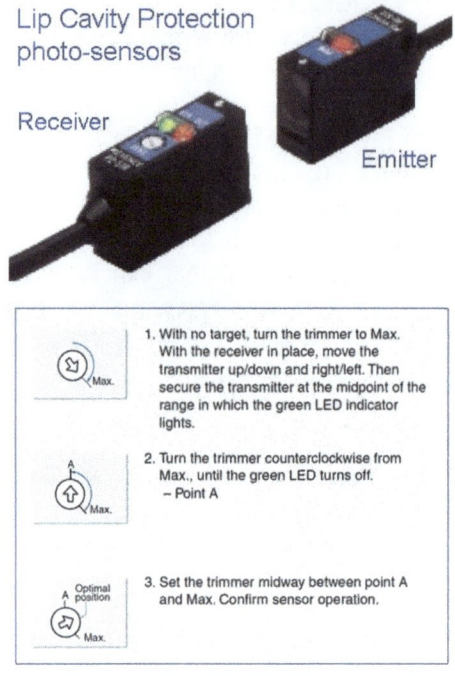

Lip Cavity Protection photo-sensors

Receiver / Emitter

1. With no target, turn the trimmer to Max. With the receiver in place, move the transmitter up/down and right/left. Then secure the transmitter at the midpoint of the range in which the green LED indicator lights.
2. Turn the trimmer counterclockwise from Max., until the green LED turns off. – Point A
3. Set the trimmer midway between point A and Max. Confirm sensor operation.

Proximity Sensors contain a *coil* in which a fluctuating magnetic field is induced by applying an AC current,. The approach of a metallic object influences the induced field, causing a change in the applied current which then triggers a signal.

Proximity Sensors provide a high-speed response compared to switches requiring physical contact. Detection takes place with almost no effect from dirt, oil, or water on the object being detected, and since they use semiconductor outputs, there are no contacts to be damaged and shorten the service life.

Photo-sensors

Two different types of photo-sensor are used in one-stage machines. A two-part (emitter and receiver) photo-sensor mounted on either side of the injection station is positioned so that if the Lip Cavities don't close correctly, an infra-red light beam is obscured, signalling to the PLC that it is not safe to issue the 'mould close' instruction. Elsewhere a retro-reflective photo-sensor "counts" the preforms as they pass before it. Should the number be incorrect, the machine cycle is stopped to allow the operator to check whether a preform has been left in a cavity.

Safety Door Switches

Should a hinged safety door be opened, safety-interlock switches mounted on the frames cut the power supply both to the motor driving the hydraulic pumps and to the solenoid-operated direction valves on the machine, thus halting all machine movements.

Safety-interlock switches may be either hinge-actuated, or operated by the insertion of a specially-

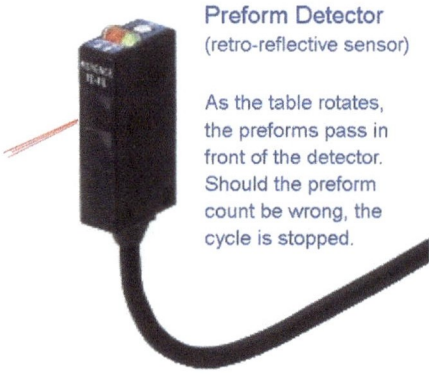

Preform Detector (retro-reflective sensor)

As the table rotates, the preforms pass in front of the detector. Should the preform count be wrong, the cycle is stopped.

profiled *tongue* mounted on the leading edge of each door. Tongues have a self-ejecting feature to prevent a non-attached one being inserted to enable the machine to operate with a safety door open (which means that doors fitted with this type of switch require a mechanism to latch the door in the shut position). It is vital that the safety-interlock switches function correctly and so the alignment of the tongue to the switch, and correct operation must be checked regularly.

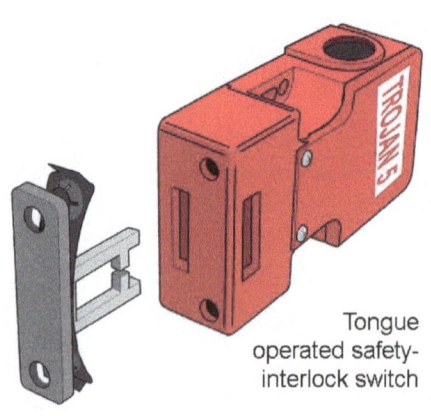

Tongue operated safety-interlock switch

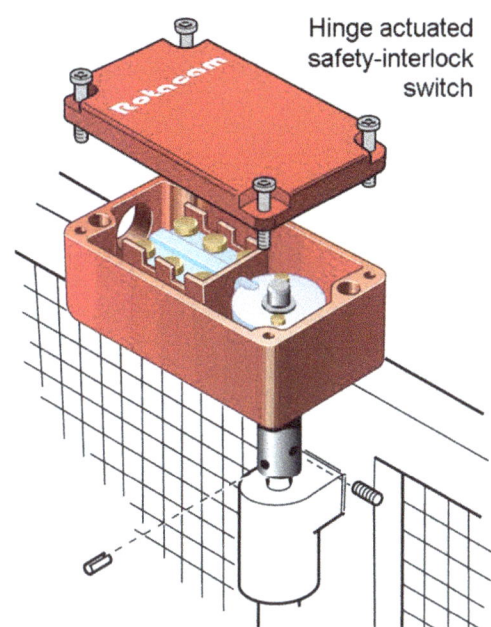
Hinge actuated safety-interlock switch

Modern safety-interlock switches are more complicated than they appear, containing multi-pole magnetic switches that provide redundancy as well as the capability to continuously monitor the integrity of each safety circuit.

Electrical schematics

Schematics use standard symbols to show how the components are connected together in a circuit. Each component symbol has a number of connection points or terminals to which lines can be drawn. These lines represent actual wires inside the machine. Wires can connect two terminals together, or they can connect several. To make schematics more legible a terminal is labelled, typically with a two or three digit number. Terminals with the same number are assumed to be connected, even though a line isn't always shown joining them.

When a wire splits into two directions, it creates a junction. Junctions are represented in schematics with nodes (dots at the intersection of the lines) identifying that the crossing wires are physically connected. Voltage nodes are symbols to which component terminals can be connected in order to assign them a specific voltage level. All terminals connected to a like-named voltage node are connected together. The ground voltage node is especially useful because so many components need a connection to electrical ground (earth).

Larger schematics are split into functional blocks on separate pages. Typically these will include sections for Control Power, Thermocouples and Heaters, Inputs (from both selector switches and proximity sensors), and Outputs to solenoid operated direction valves.

Energy efficiency

The conversion process in the manufacture of plastic containers involves melting the polymer, forming it into a desired shape and then cooling it. The action of heating and cooling entails the use of substantial amounts of electricity. However only about 10% of the energy used by an injection moulding machine is input to the polymer, with the remaining 90% used for machine functions. This means that there are a number of process adjustments that can help to reduce energy requirements:

- plasticisation set faster than required — typically the screw rotation speed should be set no faster than is necessary to ensure that screw charging is completed just before the mould is due to open.

- excessive holding pressure applied for longer than needed — once a thermal gate has frozen or a valve gate has closed, holding pressure has no effect.

- barrel and hot runner temperatures set too high — reducing the melt temperature can significantly reduce the energy requirements and also shorten the cycle time.

- compressed air is very expensive — ensure the blow air pressure is set at the minimum level for product quality, and fix any leaks promptly.

- reduce set temperatures if the machine won't be running again immediately (including the mould Temperature Controller and the Material Dryer).

- chilled water is expensive — instead of also using it to cool the machine hydraulics, install a separate cooling water system; and ensure that chilled water supply pipes are insulated.

Table of Electrical Symbols

Symbol	Description
	Crossing wires are connected
	Crossing wires are not connected
	Normally Open Single Pole/ Single Throw Switch (SPST), connects current when operated
	Normally Closed Single Pole/ Single Throw Switch (SPST), disconnects current when operated
	Normally Open Double Pole/ Double Throw Switch (DPDT), connects current when operated
	Normally Open Push Button Switch connects current when operated
	Electrical Resistance (Heater Band)
	Electric Motor
	Transformer changes AC voltage from high to low (or low to high)
	Lamp/ light bulb
	Earth Potential (ground)

8

Plant engineering

The industrial revolution has raised the quality of life for many people around the world, but it is becoming clear that it also has adverse consequences for the global environment. Despite this, demand for manufactured goods continues to grow; which means that even though manufacturing is part of the problem, it will also need to be a part of the solution.

Sustainable manufacturing seeks to meet the needs and aspirations of the present without compromising our ability to meet those of future generations. This means understanding the negative impacts of our operations and products on the environment (harmful emissions, a reliance on non-renewable resources), and working aggressively to reduce them. Good plant engineering can help here; while cutting waste by using raw materials, water, and energy more efficiently benefits not only the environment, but also the bottom line.

Factory layout

The principles of ISBM production are straightforward: polymer that enters a facility at one end is converted by automated ISBM machines and toolsets into containers which ultimately are despatched to customers from the other end. Accordingly the *type* of container to be produced (beverage, household, pharmaceutical, etc.), and its *specification* (capacity, neck size, volumes, decorated or undecorated, filled or unfilled, methods of packing and shipping) largely determine the choice of production machinery and thus the factory layout. Of course provision still needs to be made for *volume* and *product flexibility* (will the line cope with future products?), as well as access for maintenance and tool changes (a travelling crane over the machines can help make tool changes easier). Getting it wrong will lead to inefficiency and maybe excessive quantities of *work in progress*, leading to increased costs. Altering a layout can be time-consuming and expensive, so it is best to get it right the first time.

Material handling, ranging from dealing with bulk deliveries, facilitating a "first in, first out" strategy, through to drying the pellets before processing must be considered from the outset. A centralised drying system feeding multiple machines will minimize the need for manual intervention, but to prevent moisture regain it is essential that dry air is used to convey the dried pellets to the machine-mounted receivers. These (together with the distribution pipe work), should be thermally insulated and similarly fed with hot, dry air so that the pellets inside them remain both dry, and hot. Ideally the conveying line will have a valve to shut off material flow so that it may be completely emptied to avoid any pellets cooling in the line. Where feasible, situating material handling and drying equipment on a mezzanine floor ensures that floor-level access is not obstructed.

Depending upon the needs of the application, the polymer may be delivered pre-compounded (already mixed) with colorant by the supplier, or more frequently the convertor must blend a separate *masterbatch* with the pellets. For smaller volumes it may be feasible to do this manually away from the machine, but automated masterbatch dosing systems fitted at the injection cylinder throat are both more convenient and efficient. They can either be gravimetric or volumetric in design, depending upon the nature of the colorant. (Gravimetric systems more readily accommodate different materials with dissimilar pellet shapes and bulk densities).

Containers produced in small volumes may simply be packed into cartons beside the machines, but where production volumes justify it, or where downstream operations such as labelling or printing, and palletizing require it, a *bottle take-out* device can automatically transfer the containers to a conveying system. Conveying systems need to incorporate *accumulators* to temporarily store containers; so that it isn't immediately necessary to stop container production the minute a downstream breakdown occurs. Keep in mind that labelling prior to despatch (rather than immediately after blowing) can reduce inventory storage space requirements because then only bare containers and labels need be stocked (instead of stocking minimum quantities of each container having a different label).

Effective supervision is facilitated by good layout. Supervisors need to know which machines are producing and at what output rate? Correct data enables more accurate scheduling and delivery forecasting. Enterprise Resource Management (ERP) computer software is available that provides real-time manufacturing/ shop-floor performance reporting. Ideally when a new production line is specified, provision will be made for fitting the necessary sensors even when there are no immdediate plans to install such equipment.

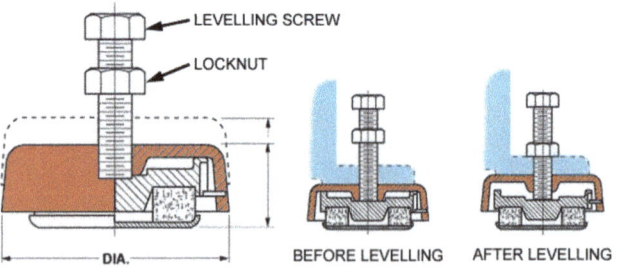

Factory floor loading

A building's *Permissible Floor Loading* is the maximum pressure expressed in kN/m^2 that the floor can safely accommodate. A competent building engineer must review the machine installation plan and verify that the factory floor is of adequate strength. ISBM machines should be set on a sound and stable steel bar-reinforced concrete slab floor. Ideally *Level Pads* will be fitted at each mounting point. These are heavy metal support housings with solid rubber vibration-absorbing feet in contact with the floor. The Pads should be attached to the machine using the levelling screws before the machine even contacts the floor. After the machine has been lowered into position, first ensure that its weight is carried by all of the Pads before levelling to the desired height by gradual and sequential adjustment of the levelling screws. (Verified using a spirit level placed on any accessible machined paint-free horizontal surface). Only once this has been completed should all the locknuts be tightened.

Factory lighting

By getting the lighting right, staff will be more comfortable in their working environment. Research indicates that daylight has tremendous physiological and psychological benefits: bringing daylight into factories (through roof lights as well as windows) can help with concentration and benefit productivity. Of course using daylight successfully within factories is not always simple or straightforward. For example direct sunlight onto bottles must be avoided lest heat gain lead to them distorting or shrinking. Artificial lighting will still be needed, but by ensuring it is linked to the daylight availability, it will be used most efficiently.

In the past lighting has consumed as much as 20% of the electricity used in industrial buildings, but this can be reduced by using modern energy efficient lighting. Make sure to choose a reputable lighting supplier and specify that an energy efficient lighting design is wanted. During procurement and commissioning ask the lighting supplier to confirm that the performance of products used achieves the levels specified in the product data sheets, and that the installation is in accordance with the design.

Determination of machine Annual Output

To determine the annual output for each machine using the chart opposite, begin by locating the actual or estimated *cycle time* along the bottom, then trace a vertical line up the chart until it intersects with the coloured curve that equates to the number of cavities, and simply read the potential annual production off the left hand scale. It is assumed that the production year consists of 8,400 hours, and no provision is made for unplanned stoppages.

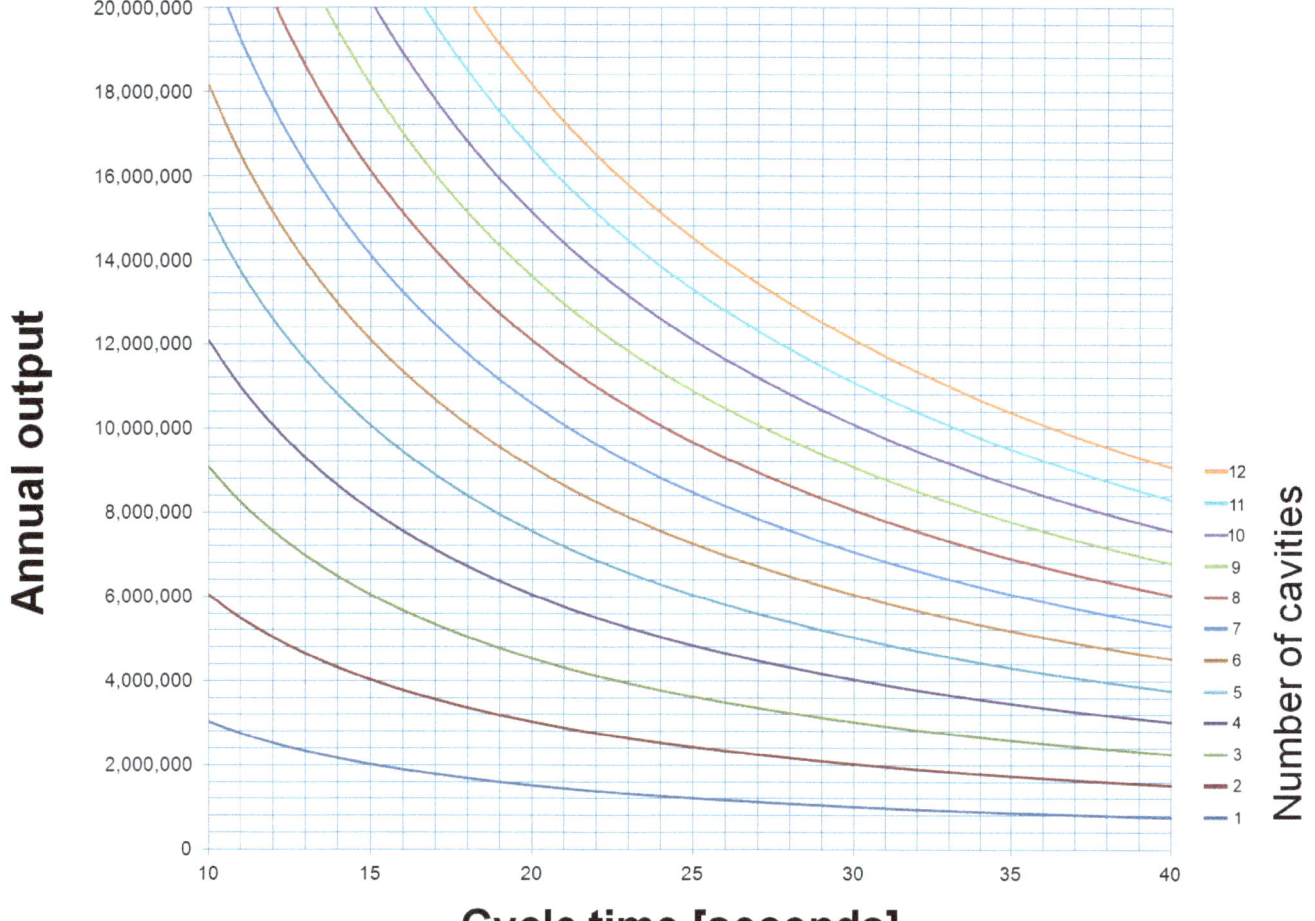

Calculating weight of polymer required

Once the annual production quantity is known, the total weight of polymer that will be needed to produce this quantity of containers is easy to determine since the *number of metric tonnes* required to make every one million containers is simply the preform *gram weight*. For example, to produce one million containers each weighing 25 grams, requires 25 tonnes of material, so to produce 600,000 or 0.6 million 25g containers would require 0.6 x 25 = 15 tonnes.

Cooling systems

The cooling capacity required by each machine will be specified by its supplier. For fast PET preform moulding times, the injection mould tool will need to be cooled with chilled water. The material dryer and the machine hydraulics also require cooling, although at somewhat warmer temperatures. What this means is that for larger installations at least, it is generally more energy efficient to install two separate cooling circuits: *chilled* water (10-15°C) and *cooling* water (20-30°C).

Remember that the key to rapid heat transfer (and shorter moulding cycles) is turbulent flow in the cooling lines. As a *rule of thumb*, turbulence can normally be achieved by having minimum flow rates of about 5 litres/min/kW of cooling capacity.

Compressed air

The quantity or volume of *operation air* an ISBM machine requires is calculated by adding together the total capacity of the internal pipe work and the various actuators, and multiplying the result both by the number of times that this volume is swept in an hour, and by the (maximum) supply pressure to the machine of 1.0 MPa. The volume of *blow air* is calculated by multiplying the capacity of the container by the number of cavities, and multiplying the result by the number of cycles per hour *and* the blow air pressure.

The total air requirement of a facility isn't as might be assumed, the sum of the maximum requirement for each process, but the sum of the *average* air consumption of each one. Oversized air compressors are extremely inefficient because when operating at part-load they consume more energy per unit volume of air produced. Plants with wide variations in air demand will need a system that operates efficiently under part-load. High short-term demands should be met by taking stored air from an air reservoir. To allow for efficient operation at times of low demand, use several smaller compressors with sequencing controls.

A compressor is sized to exceed the expected blow air requirement so that it doesn't have to run continuously. At higher altitudes or when the inlet air is hotter, it will be less dense, which means the mass flow and pressure capability of a given compressor will be reduced. In order to determine what size compressor will be needed to deliver the calculated volume of air under actual site conditions, correction factors must be applied which take into account the height above sea level, the air temperature, and the humidity (all factors that affect air density), not forgetting to add in a reserve to cater for the possible installation of additional machines in the future.

What this means is that a compressor is likely to start its life on a low *duty cycle*, but as it ages and further loads are added, it will be operating for more of the time. Leaks cause a rise in compressor duty cycle, as do blocked filters restricting the air flow into the compressor. The duty cycle of a compressor thus gives a good indication of the "health" and reserve capabilities of the system.

The two main compressor types are positive-displacement and dynamic. Dynamic compressors impart velocity energy to continuously flowing air by means of impellers rotating at very high speeds as in a centrifugal compressor. In a positive-displacement compressor air is contained in a compression chamber whose volume is mechanically reduced to cause a corresponding rise in pressure. At constant speed, the air flow remains essentially constant but the discharge pressure may vary.

initial cost, compact size, low weight, and are easy to maintain. They are energy efficient, may be air- or water-cooled, and provide pulse-free high output volume over a long life. They are commonly employed to supply the 0.96 – 0.98 MPa operation air used to operate the pneumatic actuators on blowing machines. Booster compressors are available (although not widely used) to increase the pressure of the air produced by a rotary screw compressor up to levels suitable for bottle blowing.

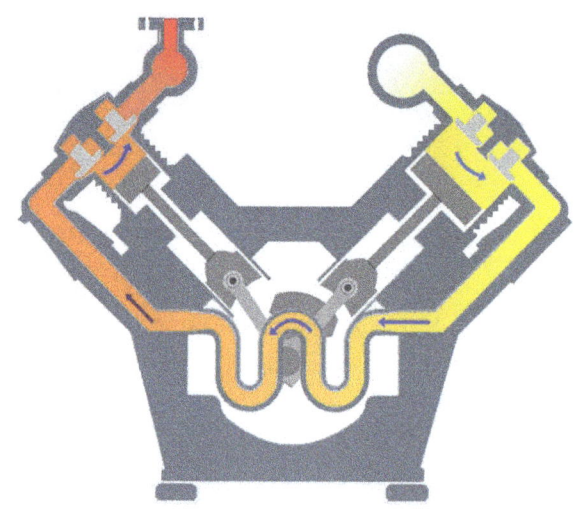

Two-stage single-acting piston compressor

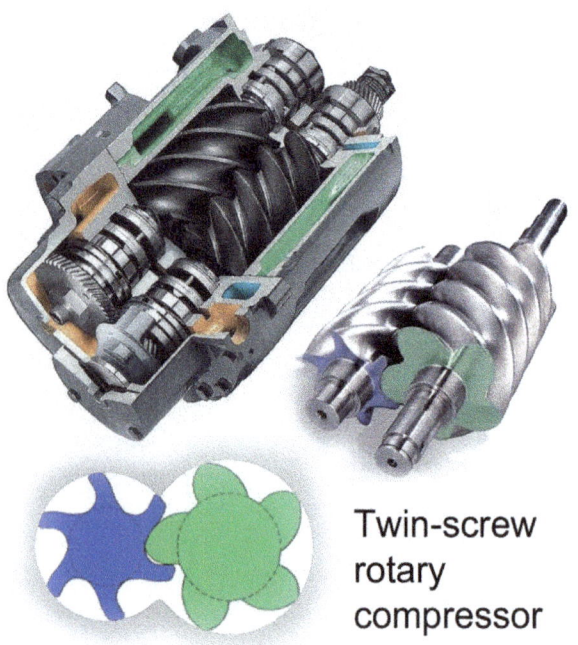

Twin-screw rotary compressor

Positive-Displacement Compressors are available in two types: *rotary* and *reciprocating*. The most common type of rotary compressor is the helical twin-screw type (also known as rotary screw or helical-lobe). Male and female screw-rotors mesh, trapping air, and as the rotors rotate to the discharge point, compressing it. Screw compressors have low

Reciprocating compressors work much like a bicycle pump. A piston, driven through a crankshaft and connecting rod by an electric motor, reduces the space inside the cylinder occupied by the air, compressing it to a higher pressure. The diagram is of a multi-stage piston compressor in which the air is compressed in two stages, first increasing the pressure (on the right-hand side) before entering into the second stage where the air is compressed to even higher pressures (in the smaller cylinder on the left).

When only one side of the piston is used for compression as shown here, it is described as single acting (as opposed to double acting when both sides of the piston, top and bottom are employed). The piston compressor is the only type capable of compressing large volumes of air to the high pressures needed for stretch-blow moulding applications. Multi-stage, double acting compressors are the most efficient compressors available, although they are typically larger, noisier, and more costly than comparable rotary units.

Sometimes the displacement of the compressor is quoted, rather than the air it delivers. The displacement is the swept volume of the cylinder(s) multiplied by the number of times they reciprocate in one minute. It is higher than the air delivered as not all of the air is pushed into the receiver owing to the back-pressure of the air already inside.

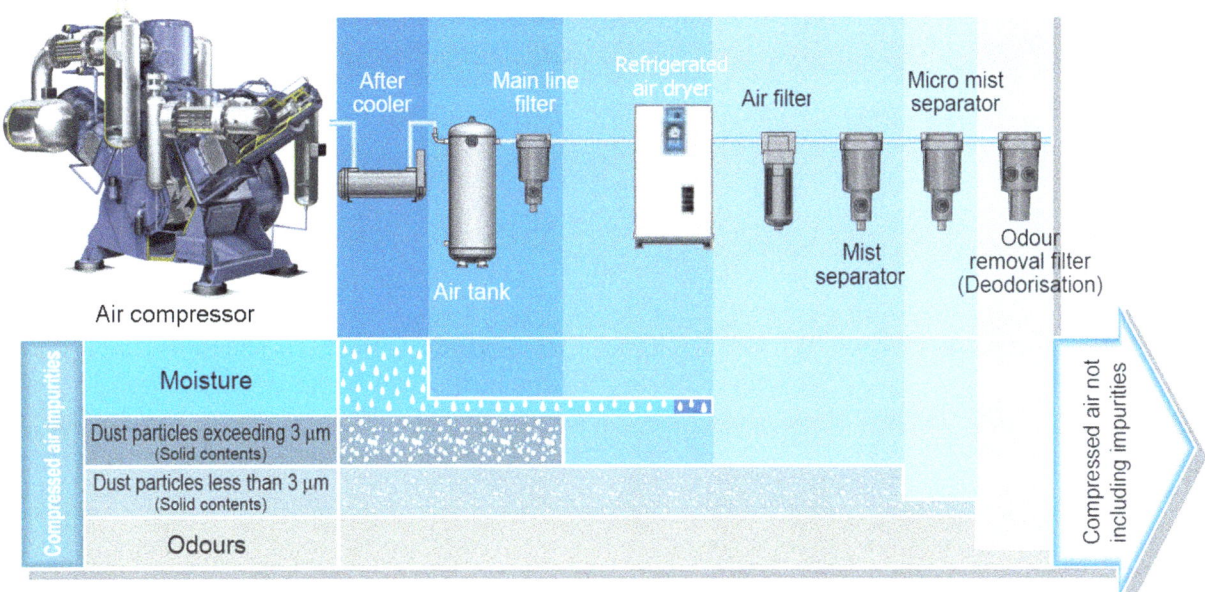

Compressing air generates heat and so single-acting reciprocating compressors are typically air-cooled by a fan blowing cooling air across finned surfaces on the outside of the compressor cylinder's cooler tubes. The cylinders of larger, double-acting reciprocating air compressors are water cooled.

Typically a compressor is used to maintain the pressure in an *air reservoir* to which the air tank inside the ISBM machine is connected. The reservoir not only stores compressed air to meet peak demand, but dampens pressure pulsations (in the case of reciprocating compressors), provides the opportunity for radiant cooling to happen, and last but not least enables condensate collection. As the air in the reservoir is consumed, its pressure falls until at a *set minimum* pressure the compressor is re-started to replenish it. The compressor then either runs for a fixed period, or is automatically stopped once the pressure in the reservoir has reached the *set maximum* level. The reservoir should be sized (in litres) to be at least 6–10 times the compressor free air output (in litres/ second), and be situated near the area of highest demand.

The atmosphere contains moisture. The higher the ambient temperature, the more moisture it holds. Compressed air contains "concentrated" water vapour which must be removed by means of a refrigerated air dryer to cool the compressed air until the vapour condenses. Subsequently these water droplets, together with entrapped lubricating oil and debris that may have been ingested, are removed by filtration. A variety of filter elements is available, including *coalescing* models capable of removing oil aerosols and particles down to 0.3 microns, and

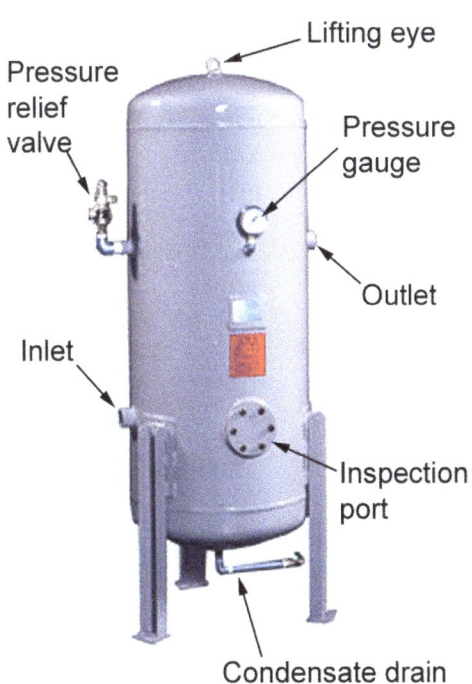

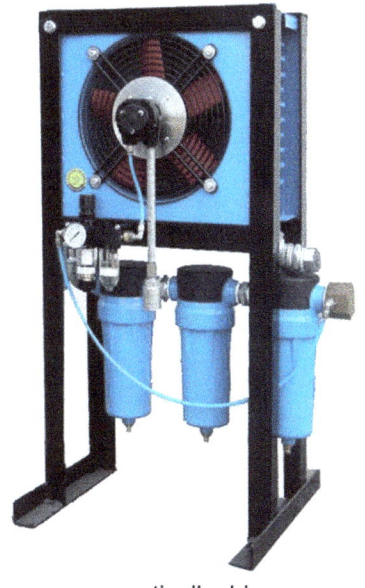

pneumatically driven atmospheric after-cooler with three-stage filtration *activated carbon* filters which are able to remove the oily smell sometimes found in compressed air.[8] Note that when a filter becomes blocked the pressure drop across it increases, reducing the pressure supplied to the machine. The cost of this loss of air, even over a short period of time, can be much greater than the cost of a replacement filter. Differential pressure gauges should be installed across filters to monitor their condition. Typically the pressure drop across a new filter should not exceed 3 psi.

A well-maintained compressed air system needs less energy to deliver the required pressure, so filter elements should be changed, and condensate collection systems checked regularly to ensure that they're working effectively. Compressor efficiency is determined largely by the condition of valves, piston rings and similar components subject to friction wear. These should be examined at the intervals stated in manufacturers' instruction manuals and replaced when necessary.

When designing a compressed air distribution system it is best to locate the air compressor and ancillary equipment where the temperature inside the plant is low (but not freezing). Take into account that for every 30 metres of supply pipe one can expect to lose 6 kPa pressure, and that valves, elbows and T-sections in the line also bring about pressure losses. As with machinery layouts, wherever possible provision should be made for a possible future increase in the number of ISBM machines. Modern CAD software enables rapid creation of automatically piped 3D plant concepts that can be evaluated for maintenance and operability concerns well in advance of detailing.

Example: calculating blow air requirement

Determination of the maximum volume of blow air required by an ASB-70DPH in Normal cubic metres per hour (Nm^3/Hr), i.e. the volume that the air would occupy at 0°C, 0% RH, and at sea level.

Nm^3/Hr = ((*OFC* x *Cavities*) + *Piping*) x 60 / *Cycle Time* x (P + 1) x 1.3

OFC	=	Overflow capacity of container (litres)
Cavities	=	Number of blow cavities
Piping	=	Capacity of blow air pipes on the machine (litres)
P	=	Blow pressure (kgf/cm²)
1.3	=	safety factor allowing for a loss of efficiency over time and air leakage

For reference, the capacity of the blow air pipes on a Nissei ASB-70DPH machine is taken to be

---

[8] Compressed air quality levels are defined in ISO 8573-1:2001 Compressed air - Part 1: Contaminants and purity classes

3.6 litres. (This figure may appear unfeasibly large at first sight, but remember that we're determining the *maximum* required volume, and since the standard ASB-70DPH machine is designed to accommodate up to an 8-cavity toolset, a considerable length of air hose will be required).

The ASB-70DPH machine is capable of producing a wide variety of containers in different volumes and quantities. The combination resulting in the largest "blown volume" is **four x 3 litre** containers. Assuming a cycle time of **20 seconds**, and based upon the maximum permissible blow air pressure of **35 kgf/cm²**, inserting these figures into the above equation gives:

Max volume of blow air
= ((3 x 4) + 3.6) x 60 / 20 x (35 + 1) x 1.3

= 2190.24 Normal litres/min

divide by 1,000 [litres in 1 m^3] and multiply by 60 [minutes in one hour] to obtain Nm^3/hour

= **131.4 Nm^3/Hour**

Free Air Delivery

The free air delivery (FAD) is the volume of air sucked in under the conditions of temperature, pressure, and humidity prevailing at the compressor inlet. It is calculated by applying correction factors to the Normal cubic metres per hour figure.

Example: taking the same ASB-70DPH, if the facility was situated 300 metres above sea level and the extreme ambient conditions were 40°C, 80% RH, then reading from the graphs:

Temperature & Humidity Correction: at 40°C and 80% RH the correction factor is **1.23**

Altitude correction factor: at 300m above sea level the correction factor is **1.04**

Then the size of compressor required to deliver 131.4 Nm^3/Hr is determined by multiplying this figure by <u>both</u> correction factors:

131.4 x 1.23 x 1.04   = **168 m^3/Hr FAD**

Use the Combined Gas Law to calculate the volume of this air when it has been compressed:

$$\frac{P_1 \times V_1}{T_1} = \frac{P_2 \times V_2}{T_2}$$

Temperature & Humidity correction

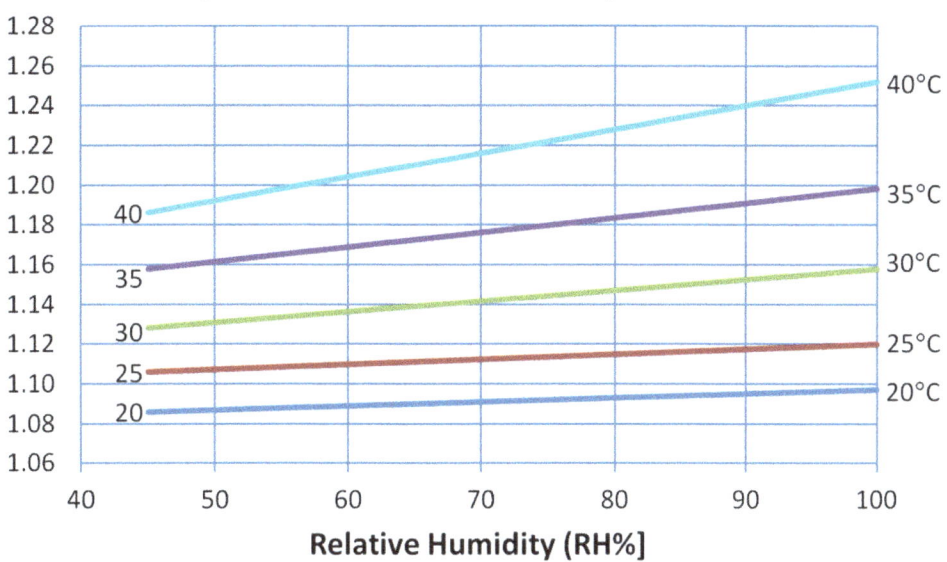

Altitude correction factor

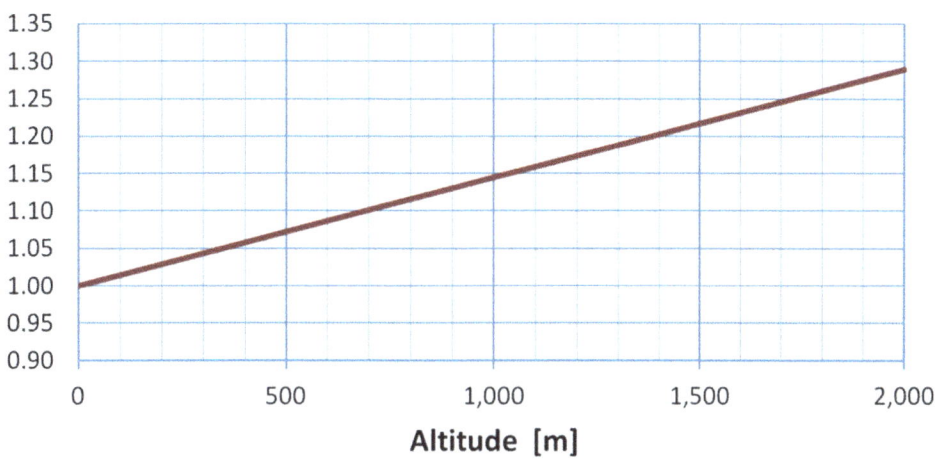

Note that both temperatures must be in *Kelvin*, (obtained by adding 273.15 to the temperature in °C). Rearranging the equation to give V_2:

$$V_2 = \frac{V_1 \times P_1 \times T_2}{T_1 \times P_2}$$

Suffix 1 relates to the inlet conditions, and suffix 2 to the outlet conditions. So V_1 is the calculated Free Air Delivered, and V_2 is the volume discharged from the compressor outlet at temperature T_2 and pressure P_2.

It can be difficult to visualise the difference in volume of air before it has been compressed (i.e. Free Air Delivery) and after it has been compressed. This image offers a "cartoon" representation of the difference:

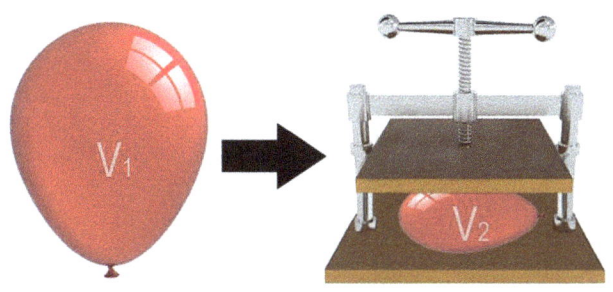

Air volume before compression (FAD)   Air volume after compression

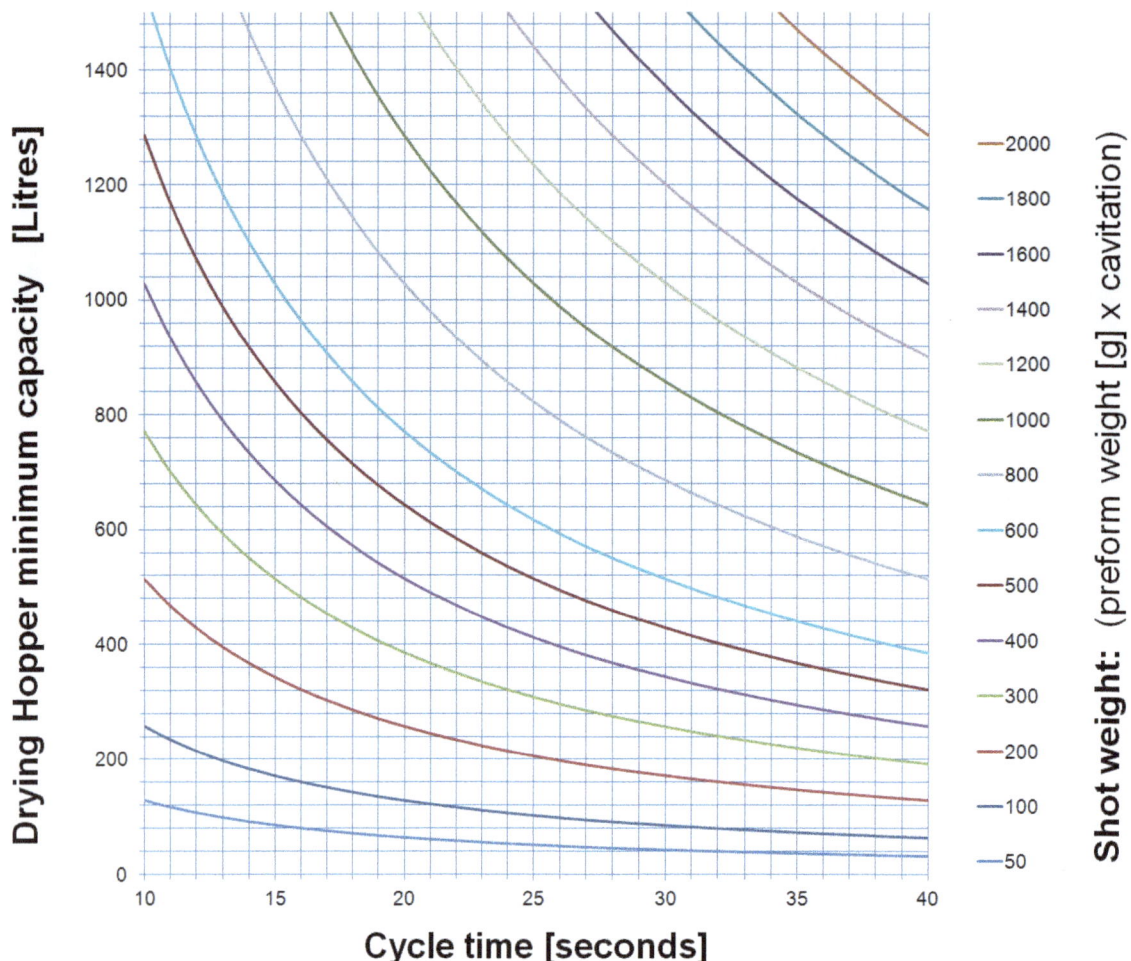

Material Drying

It is crucial that PET resin is dried at the correct temperature and time. Ideally this will be six hours (certainly not less than four hours), which means the drying hopper must be large enough to hold four to six hour's worth of resin. Not only will insulating the hopper save considerable amounts of energy, but the resin close to the walls of the hopper is unlikely to reach the proper drying temperature without adequate insulation. The hopper design influences both the heat distribution inside it, and the material flow through it. *Funnel flow* occurs when the pellets flow faster at the center of the hopper than they do around the side. A tall slim hopper design helps avoid this, thus ensuring all the pellets are exposed to the hot drying air for the same length of time.

Use the chart above to determine the minimum hopper capacity that will be needed when producing a particular PET container. Having first calculated the *shot weight* in grams (by multiplying the preform weight by the number of cavities), locate the *cycle time* along the bottom axis, then trace a vertical line up the chart until it intersects with the coloured curve nearest to the shot weight, and simply read the minimum hopper capacity in litres off the left hand scale. It is assumed that the shot weight is in grams of PET, the cycle time is in seconds, and the drying time will be six hours.

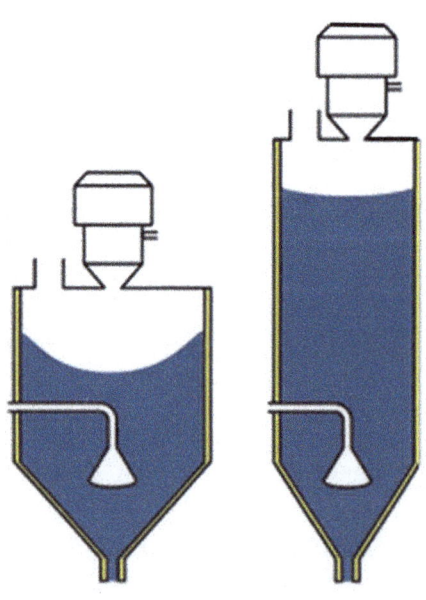

"Funnel flow" is less likely to occur in the tall slim hopper on the right giving more effective drying.

Chapter 8—Plant engineering 87

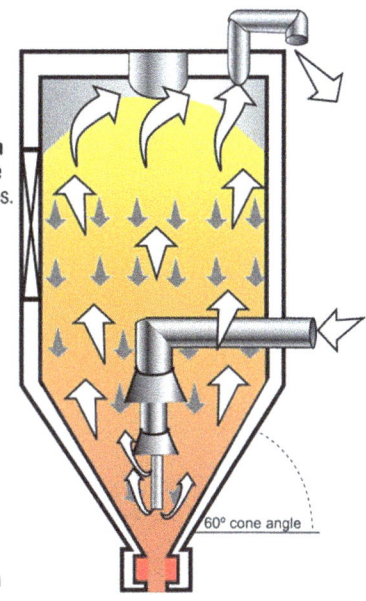

Extra insulation in the side walls maintains temperature at desired level, prevents heat loss and protects workers from hot surfaces.

A large access door and smooth interior walls minimize downtime for clean-out and material changes.

Mass material flow Smooth surfaces and steep cone angles ensure that each pellet is exposed to heated drying air for the specified drying time.

Even air/heat distribution throughout the hopper. The spreader improves air distribution and provides better process air to resin contact throughout the hopper.

Magnet to attract and securely retain any ferrous contamination

Daily machine maintenance

The effectiveness of a production line clearly depends on many factors, but in the final analysis, if a machine has broken down, its rated output or versatility is irrelevant. Which is why the daily maintenance regimes recommended by machine manufacturers are so important (as is establishing a recording system to verify that they have actually been carried out). Regular cleaning of an ISBM machine (inside the guards, as well as outside) might not seem of the highest importance, but the intimate knowledge it affords of gradual changes in the machine's condition can be crucial.

Hand tools and replacement parts stock

Relatively few hand tools are required for a typical toolchange. These include a selection of both short and long ball-ended Allen (hex) keys for the fixing bolts, a short steel pipe to use as a lever when *undoing* hex bolts, the appropriate open-ended spanners for the water and air hoses, a mallet with plastic striking faces (to avoid inflicting damage to the tool), eye bolts, shackles, and slings for lifting. Ideally these will all be kept together in one toolbox.

A spare parts store situated near the machines is essential; as is a strict policy of recording which parts have been used. The latter will be useful not only when reordering spares, but also to help identify recurring machine problems. The stock needs to be properly organised so that parts can be found quickly and easily when needed. It should include ample supplies of NOMEX papers, replacement bolts and O-rings, as well as carefully chosen machine parts. Holding manufacturers' recommended spare parts in stock might seem an expensive luxury, but generally proves cost-effective over the longer term.

Toolset maintenance

When moulds are mounted in a machine, it is important to avoid corrosion by initially coating the platens with a film of oil or grease or spraying them with a proprietary anti-rust product. Equally when moulds are removed from the machine it is vital to prevent the highly polished moulding surfaces from corroding, and so they need to be protected with an anti-rust product. Purpose-designed spray products displace any surface moisture, and then rather than "running off", form a viscous protective film that protects the steel against corrosion during storage. Some incorporate a coloured dye which allows easy visual identification to ensure complete coverage.

Cleaning moulds is a key part of running a successful moulding operation, either on or off the machine. Cleaning solvents have different degrees of effectiveness depending upon the environment. Solvents that work well at room temperature will likely evaporate very rapidly from a hot mould and so won't be in contact long enough to do much cleaning. Solvents that work well at higher temperatures are much less aggressive and effective at room temperature. They take longer to evaporate and may even have to be wiped off. As a general rule, cleaning solvents used on a warm or hot mould should be much slower evaporating to allow for the chemicals to remain on the mould surface for longer periods of time, thus maximizing their cleaning potential.

Mould polishing compounds are typically used in the form of a paste containing a combination of cleaning solvents and mild abrasives. They require manual application and removal and so the provision of a plentiful supply of cloths or paper towels is essential.

Moulds put into long-term storage will ideally will be packed in some form of stackable, pallet-mounted container with a secure lid. Also the fixing bolts should be kept with the toolset in a properly labelled container.

"folding" steel pallet box for tool storage

9

Quality Assurance

Quality involves much more than a simple check before despatch that a product complies with its specification. Ideally containers are inspected and tested not only to confirm compliance, but also to obtain information about the production process. A variety of systems to assure quality have been devised. The more widely used are ISO 9000 (which is mainly procedural), Total Quality Management (essentially cultural), and Statistical Process Control (which as the name suggests uses statistical techniques to minimize the number of containers that need be measured).

Whichever system is chosen, in the final analysis it will be necessary to measure container attributes that are critical to quality. Product tests should be straightforward to perform, objective, not subjective, and selected with half an eye on how the bottle is to be used by the consumer. There is no value in measuring something that doesn't affect performance, aesthetics, or is irrelevant to the way the container is used. Test results need to be retained in case proof of compliance is needed in the future, and the data recording method should also lend itself to highlighting process trends (which means for example that cavity numbers should also be recorded). Then if it happens that containers from certain cavities are always worse than the ones from the others, at the very least resources can be focused on the offenders to determine why.

In general terms the more common tests can be divided into two groups (albeit with some overlap). The tests in the first group verify a mould has been manufactured within tolerances and is correctly mounted in the machine. These tests include preform weight, neck thread form and dimensional accuracy, and bottle dimensions and capacity. Such tests are typically performed when the tool is new, but infrequently after that. The second group of tests check whether the bottles being produced are all the same, i.e. the process is under control. These tests measure material distribution (weight and wall thickness) and bottle performance (drop, top load, seal, and burst tests). These tests are performed on representative sample bottles every shift. They can be further subdivided between the quicker checks performed by the machine operator, and the more involved tests undertaken by the QA personnel.

Many different test methods have been devised for evaluating containers; those described in this Chapter have been selected as being of general application, although not all of them will be applicable to every container type, while additional tests may be required for special purpose containers. How frequently a particular test should be done will depend both on its complexity (how difficult, time-consuming and expensive it is to do), and how useful it is in providing information as to whether the process is under control.

- With the more subjective tests it can be useful to keep representative sample containers and preforms of both acceptable and unacceptable quality for comparison with current production.
- As both preforms and bottles will shrink a very small amount with time, to permit valid comparisons all preform measurements should be made 24 hours after moulding; while measurements on bottles should be made only after the bottles have been stored for 72 hours in controlled conditions.
- It can be difficult to judge the worth of any process changes when only relatively insensitive attribute data is available. The extra cost associated with obtaining data in the form of numerical measurements is likely to be trivial in comparison to the savings achieved by reducing process variability.

When it comes to assessing raw materials, generally converters don't have the means to assess the quality and consistency of the incoming polymer and so have to depend upon a supplier's reputation. Material Data Sheets often include information that is not immediately relevant. Important parameters include:

Intrinsic Viscosity: (PET)	a function of the molecular chain length and thus also of the properties of the PET polymer
Melt Flow Index: (PP)	a measure of the ease of flow of PP melt; the larger the number, the easier the melt will flow
Bulk Density:	the density of the *poured* pellets; essential when calculating the capacity of the material hopper
Fines percentage:	the proportion of pellet "fragments" mixed in with them

7.1 Preform examination

Checks on appearance and shape, weight, neck dimensions, wall thickness and polarised light inspection should be made on preforms taken from each set of lip cavities. When moulding PET, intrinsic viscosity (IV) needs to be measured in only one sample from each injection cavity (although two or even three replicate samples may be required for greater accuracy).

7.1. Appearance and shape

Minor imperfections are likely to be present in every preform, so the specification should quantify or set limits on what is acceptable. Table 7.1 may be used as a guide.

7.1.2 Preform weight

Preform weight variation or fluctuation implies that the cavities are not filling evenly. This will affect container wall thickness (and thus the top load and vacuum strength) and so weights should be checked regularly using an electronic balance.

a) Is the weight of each preform within the specified range? Typically the tolerance is ±1% of the preform weight. Small variations in preform weight may be due to mould design, mould machining tolerances, flash, short shots, etc.

b) Is the cavity-to-cavity weight variation (on a multi-cavity mould) within the specified range? Typically the tolerance is ±0.5% of the preform weight. If the variation is more than this check the preforms for sink marks, short shots or flashing, also check the gate balance and make sure the air vents are clean. Having confirmed these items, check the condition of the mould and finally the mould dimensions.

c) Is the shot-to-shot weight fluctuation within the specified range? The fluctuation in weight of consecutive preforms from the same cavity (i.e. between one shot and the next) should be less than ±0.5% of the preform weight. If the fluctuation is more than this check the preforms for sink marks, short shots or flashing and check that the moulding conditions are correct.

Table 7.1

Description	Typical Limits
Mismatch of mould parts	0.05 mm maximum 'step' height
Flash	0.05 mm maximum
Crystalline haze / streaks (PET only)	Not generally acceptable, although a lace-like collar of crystallinity within a ⌀12.5mm 'disc' centred on the gate will not usually impair container performance
Short shots / sink marks	Normally the small sink marks often found in the thicker parts of the neck such as the pilfer proof or roll-out band are considered to be acceptable
Bubbles	Not acceptable
Grease/ surface or embedded dirt	"
Crystalline slugs	"
Gate stringing	"

Table 7.2

Fault	Probable Cause
Cloudy appearance	Blow mould requires polishing
Crystalline haze	Preform too hot
Pearlescence	Preform too cold, wrong material distribution
Scratches from the blow mould	Damaged blow mould, wrong blow timing, wrong blowing rate or stretch rod speed or inadequate exhaust)
Yellow coloration	Over drying

7.1.3 Neck dimensions

Preform neck dimensions should be within the tolerances specified on the drawing. If a dimension appears to be wrong, the following should be checked first:

- was the measurement correct?
- is the preform appearance abnormal?
- is there evidence of flash or a short shot?
- in the case of diameter measurements, does the result differ according to the direction of measurement (i.e. parallel or perpendicular to the parting line)?
- do the minimum and maximum preform dimensions vary according to the injection and blow cavities or according to the lip cavity?

Having confirmed the above items, the moulding conditions, then the condition of the mould and finally the mould dimensions should be checked.

7.1.4 Uneven wall thickness

The difference between the minimum and maximum wall thickness measured near the preform gate (preform eccentricity) should be no more than 0.1 mm for preforms of up to 120 mm in length (and *pro rata* for longer preforms).

Possible causes of uneven wall thickness include:

- contamination on mould surfaces
- nozzle touch force is too strong
- mould misalignment
- injection core pin is bent
- injection core pin taper and lip cavity taper are not matched and not heavily contacting
- lip cavity taper and injection cavity taper are not matched and not heavily contacting
- injection core and lip cavity flat contact surfaces are not in heavy contact
- lip cavity manufacturing process was faulty (centres not aligned)
- injection core pin manufacturing process was faulty (centres not aligned)

When determining the cause of uneven wall thickness, it should be considered whether the minimum and maximum wall thickness and their relative positions remain the same from shot to shot (if they change it would indicate that the mould parts are able to move during injection).

7.1.5 Polarised light inspection

Examination of a PET preform under polarised light gives an indication of the stresses set up in the preform during injection moulding and curing. A properly made preform will generally have a regular stress pattern while a random stress pattern could indicate a moulding problem. Unfortunately interpreting stress patterns is difficult and so this test is not commonly used. (One possible exception being the detection of *moisture rings* due to condensation on the injection core pins).

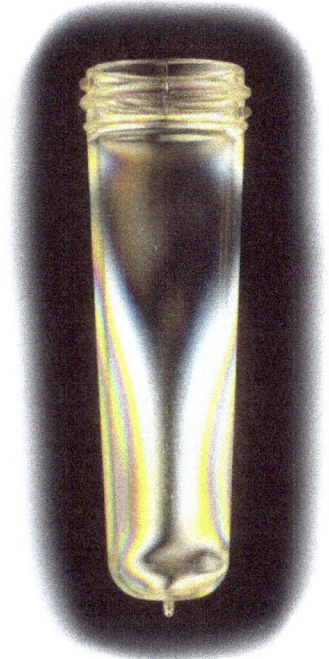

7.1.6 Intrinsic viscosity (IV)

Measuring the IV of a PET preform and comparing it with that of the virgin material can give an indication of dryer effectiveness. PET in the molten state is very susceptible to hydrolytic degradation and therefore must be correctly dried before processing. Any moisture remaining in the polymer chip will lead to a drop in viscosity during processing, resulting in reduced physical properties such as top load and increased creep in the finished container. Experiment has shown that an IV drop of 0.01 will occur for every 16 ppm of moisture retained in the PET on melting. IV testing is normally available from the PET supplier.

7.2 Container examination

Containers will shrink after moulding by an amount that will depend on the ambient conditions.

Samples should be stored and measured under controlled conditions so that test results may be compared. Containers from each blow cavity and pair of lip cavities should be tested.

7.2.1 Shape and appearance

The visual examination of a container is an essential first part of any testing program. Containers from each blow cavity and each pair of lip cavities should be examined for colour, clarity, process faults and shape definition (especially of corners or 'feet' and freedom from dents in flat faces). Common faults and their probable cause are listed in Table 7.2.

7.2.2 Dimensions

Container dimensions should be within the tolerances specified on the drawing (typically ±1mm on diameter and ±1.5 mm on height for a 2 litre bottle). Many factors can affect the container dimensions but in general, if the preform is fully blown, the container dimensions should be correct. Diameters can be quickly measured with a vernier PI tape, while 'go' / 'no go' gauges simplify height measurements. Container dimensions are affected by:

Height:
- position of bottom mould
- material distribution
- blow-mould temperature
- preform temperature
- short shot
- blowing pressure (petaloid style bottles)

Diameter:
- blow-mould opening during blowing
- material distribution
- blow mould temperature
- preform temperature

7.2.3 Capacity

The container capacity should be within the tolerance specified on the drawing. If the shape and dimensions are accurate but the bottle capacity is wrong there may be a problem with the mould design or manufacture. The factors affecting container capacity are the same as listed above for dimensions (Section 7.2.2) plus any effects due to the shape of the container, (such as rectangular or oval containers with large panels which bulge outward due to the weight of the contents, or the reduction in capacity that occurs if a container is dented).

Container capacity is affected by storage time and environment. Shrinkage of a typical PET bottle is 0.5% on volume after 72 hours storage, rising to a maximum of about 1.5% after 150 days at 22°C, (higher temperatures will increase shrinkage rates but not the peak values). Table 7.3 shows industry-standard tolerances for bottle capacity.

7.2.4 Container wall thickness and material distribution

Container specifications will often define the minimum wall thickness in each part of the bottle. When a bottle is blown it is crucial that the material is consistently and accurately placed in the correct position since a variable wall thickness will result in poorer top load, and in the case of carbonated soft drink bottles, increased carbonation loss and in extreme cases, distortion of the bottle caused by the carbonation pressure. Uneven material distribution may be due to the factors listed in Table 7.4.

Very accurate (although time-consuming) thickness measurements can be done on sections cut from a bottle wall using a micrometer. For large quantities of containers automatic measuring equipment is commercially available which can measure the wall thickness without damaging the bottle. Generally however, a check on the average material distribution is all that is required. One method is to cut the bottle into sections and compare the sectional weights with those of sample containers previously found to have acceptable material distribution. An electrically heated hot-wire cutter can provide the precision cutting required.

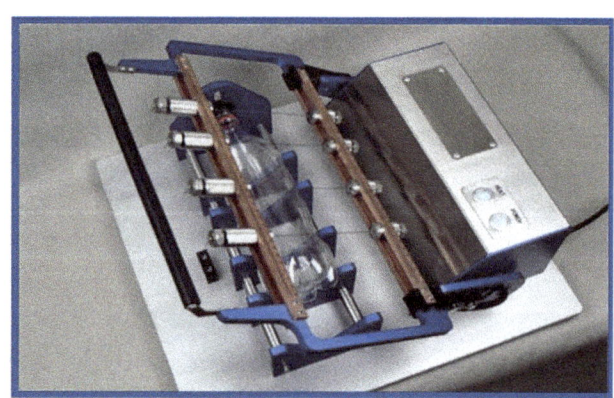

7.2.5 Top load strength

Containers are required to withstand the high top loading forces imposed by the filling and capping processes and during stacking. Top load strength is a function of a container's shape and wall thickness. Testing involves subjecting the container to an increasing load applied vertically down on the neck. The top load being defined as the applied load when the container starts to collapse. For the container to be considered acceptable, this load should be greater than the maximum load to which it will be subjected during its life. Typically top load strength is measured on an empty, uncapped container under standard conditions (Table 7.5).

During production, the top load strength of bottles from each blow cavity should be measured regularly (e.g. twice per shift), and the results recorded and analysed since a reduced top load may indicate production problems such as inadequate drying (which causes an increased viscosity drop in processing, leading to wall thickness variability).

7.2.6 Impact resistance (drop) test

Ideally, water-filled and capped containers should not break when dropped accidentally. Drop testing for carbonated soft drink bottles is typically carried out as follows:

- two sets of test bottles are filled with carbonated water and stored for 24 h at 24°C and 4°C respectively
- a bottle, held at an angle of approximately 30° from the vertical and 2.0 m above a concrete surface, is released and allowed to fall freely
- the bottle is inspected—to pass the test there should be no leakage and the bottle should be able to stand upright unaided

Acceptable drop impact performance is important for any container, but particularly so for baby feeding bottles which are often made from PP. These bottles may be stored in a refrigerator, and PP becomes brittle at low temperature. In this case the test would start with sample bottles being conditioned in a refrigerator for a sufficient length of time for both the bottle and its contents (typically water) to cool to the specified temperature, e.g. 5°C (the standard domestic refrigerator temperature), before they were dropped from a specified height onto a hard surface.

To make any drop impact test more rigorous, first identify the most critical impact point(s) and then devise a test method that will ensure test samples always impact on that critical point (e.g. by "sliding" them down vertical guide rails or tubing). The drop height is also important and should be chosen to replicate expected usage.

7.2.7 Leakage of liquid (Seal integrity)

The seal integrity of carbonated soft drink bottles is very important for the retention of the CO_2 pressure. Fortunately, because the neck finish of a stretch blow-moulded container is injection moulded, leakage problems are very uncommon. If they do

Table 7.3　　Typical bottle capacity tolerances as derived by an international soft drink company

Bottle capacity (ml)	Individual tolerance (as delivered)	Individual tolerance (as manufactured)	Arithmetic average (as manufactured)
330-500	+8, -11	+8, -6	+2.5, -0
501-1000	+9.5, -14.5	+9.5, -4.5	+5, -0
1001-1250	+11, -19	+11, -6	+6, -0
1251-1500	+13, -22	+13, -7	+6, -0
1501-2000	+17, -28	+17, -9	+8, -0
2001-3000	+25, -28	+25, -13	+12, -0

Table 7.4　Common causes of uneven material distribution in a bottle

Circumferential	Vertical
container shape (oval designs)	container shape (waisted designs)
uneven wall thickness of preform	uneven gate balance
wrong temperature distribution in preform	wrong conditioning temperatures
conditioning parts misalignment	wrong stretch speed
wrong stretch ratio or grade of PET resin	wrong primary and secondary blow timing
	wrong stretch ratio or grade of PET resin

Table 7.5　Standard conditions for measurement of top load strength

Test temperature	18-25°C
Top load application speed	50 mm/min.
Measurement point	first peak value (load at initial deformation point)

arise they are normally due to a poor cap sealing surface, the wrong cap application torque, a poor bottle sealing surface (short shot or weld line across sealing surface), a poor fit between the bottle and cap or incorrect cap material. Automatic in-line leak testing machines are available to match any line speed.

7.2.8 Vacuum strength

The cooling of hot-filled products, or the absorption of headspace oxygen by oil-containing products, leads to a pressure reduction inside the container that can cause the label panel to buckle or 'panel.' Bottle shape and wall thickness contribute to vacuum strength which can be measured by lowering the pressure inside an empty container and recording the pressure at which collapse first occurs. The test temperature and the rate of pressure reduction both influence the vacuum strength.

7.2.9 Acetaldehyde (AA)

Acetaldehyde (AA) is formed by the thermal degradation of PET during melt processing and will therefore be present in the preform and thus the bottle wall. AA in the bottle wall will diffuse into the contents and may cause a taint problem, particularly with very bland products such as mineral water. Thus it is very important that the injection cycle is set up with the optimum processing conditions to minimise the quantity of AA generated.

The most common method of comparing AA levels is by analysing the quantity that accumulates inside a sealed bottle over a set period. Immediately after moulding the bottle is purged with nitrogen gas, capped and stored at a temperature of 22°C (71.6° Fahrenheit). After 24 hours the amount of AA that has diffused into the bottle is measured using gas chromatography. For ease of comparison between bottles of different capacity, the measurement is then quoted in micrograms per litre ($\mu g/l$). How much AA can be permitted in a particular bottle will depend upon the type of beverage and how sensitive it is to tainting. Typically bottles that are intended for cola products may impart no more than $3\mu/l$ of AA to the contents, while for mineral water, levels as low as $1\mu g/l$ are required.

It is possible to measure the amount of AA in the preform but as this requires the preform to be ground up under cryogenic conditions to prevent the generation of additional AA, the test is not very common.

7.2.10 Oxygen permeation

Many food and some beverage products require the exclusion of oxygen to keep them in good condition. The amount of oxygen that permeates into a container depends upon its wall thickness and surface area to volume ratio. Machines are commercially available to measure the oxygen transmission rate through pieces cut from a bottle wall. The permeation into a container can be estimated from this provided the material distribution is known. Normally, however, these machines are modified to allow them to directly measure the permeation into a complete package. The units of permeation are $cm^2.mm / m^2.24\ h / atmosphere$.

7.2.11 Moisture vapour transmission rate

The moisture vapour transmission rate (MVTR) is a measure of how fast moisture will pass through a piece cut from a bottle wall, and like oxygen transmission, is a function of the thickness of the sample. Usually the moisture permeability of the complete pack is more important. This can be measured by filling the container with dry silica gel and storing it in a temperature and humidity controlled environment. The increase in weight due to moisture ingress can then be monitored over a period of time. The units are $g.mm / m^2.24\ h$.

7.3 Additional tests for pressurized containers

Bottles designed for containing carbonated beverages must allow for the fact that PET is a visco-elastic material that will distort when pressurised, the rate of distortion being a function of time, temperature and the level of applied stress. The elastic part of the deformation happens almost immediately and is recoverable when the pressure is released but the remainder, which takes place over a period of time, is permanent. Such deformation must be tightly controlled to prevent excessive fill level drops, an "apparent" loss of carbonation due to the increased headspace volume, and unacceptable dimensional changes.

7.3.1 Burst pressure

There are different reasons for hydrostatic pressure testing a container. In certain applications the filling process can subject the container to shock loading. Pressure testing verifies that it will survive such stresses without failing. Carbonated beverages subject their container to considerable pressure all the time, particularly at elevated temperatures. Volatile substances can have the same effect. Pressure testing verifies that a container is capable of withstanding a higher pressure than it could possibly be subjected to in use without leakage. The Burst Pressure is the pressure needed to burst or split a container. The bottle is hydraulically pressurised until it bursts in some form of test apparatus (preferably a proprietary test unit especially designed to withstand a bottle bursting). A simple pass/ fail test criteria may be easy to administer, but doesn't necessarily provide all the data that it could. Variable data arising from measurement (e.g. the actual burst pressure) is usually more informative than attribute data (e.g. a simple pass or fail). Safety procedures should be observed, especially protection of ears and eyes.

7.3.2 Thermal stability

Creep is the expansion of the bottle due to the carbonation pressure. A high level of creep would lead to an unacceptable fall in fill-height while variable wall thickness could lead to non-uniform creep; that is, distortion of the bottle shape. Well-oriented and uniformly distributed material in the bottle wall is the prime factor in minimising creep.

To test for creep, bottle height, diameter and fill point are measured before and after carbonation and storage under controlled conditions, and the percentage change determined. After the test the bottles should still be able to stand upright unaided and should not have developed a poor appearance.

The creep test may be performed in conjunction with the carbonation loss test described below, but usually an elevated temperature accelerated test is performed in which the bottle is stored for 24 h at 38°C (100° Fahrenheit).

7.3.3 Carbonation retention

PET bottles were originally developed for the packaging of carbonated soft drinks because they have the strength to withstand the carbonation pressure without undue distortion and because PET provides a reasonable barrier to carbon dioxide (CO_2). Nevertheless, CO_2 gas does slowly permeate through the bottle wall, resulting in a loss of carbonation pressure. (Note that there is also a reduction in pressure as a result of the volume increase that takes place as the bottle creeps but this is normally off-set by slightly over-pressurising the bottles on filling).

To test for carbonation loss bottles are filled with carbonated water and stored at a closely controlled constant temperature. The pressure in the bottles is monitored regularly for a set period of time determined by the bottler.

7.4 Additional tests for hot fill containers

Additional tests are required for hot fill containers to determine whether the two major hurdles to hot filling (volumetric shrinkage and leakage at the neck) have been overcome. PET containers shrink when heated because the biaxially oriented molecular chains of which they are composed have, in effect, been frozen in an unstable state and the heating provides sufficient energy for the chains to begin reverting to their original randomly folded state. However the very small crystallites present in the oriented wall anchor the molecules in their stretched positions, so the container cannot actually turn back into a preform, (although distortion of the bottle and a significant capacity reduction will occur). Secondly, the amorphous PET neck will soften at temperatures around the T_g, allowing the product to leak out, or contaminants to enter the bottle.

The various *heat setting* techniques for improving the elevated temperature performance of PET bottles essentially involve increasing the number of very small crystallites present in the oriented material by maintaining it at an elevated temperature for a period after blowing, and even on occasion, by crystallising the necks (crystalline PET does not soften until its temperature nears the melting point). A test has been developed to check the efficiency of this process for producing hot-fillable bottles (Table 7.6).

Table 7.6 Test for efficiency of heat setting

Time (s)	Action
0	Fill container to the brim with hot water at product fill temperature and leave to stand
60	Secure closure on bottle with correct torque and lay the pack on its side to sterilize inside the neck and closure
120	Stand the container upright
240	Place the pack under a cold water spray until it has cooled to room temperature
—	Determine percentage change in overflow capacity, overall height and diameter and neck dimensions
—	Check for neck leakage

Table 7.7   Test to confirm resistance to environmental stress crazing

Time (min)	Action
15	Submerge empty bottles in a 2.5% solution of NaOH (Sodium Hydroxide) in water at 59 ± 1°C
—	Remove from the solution and allow to cool naturally to room temperature (21 ± 5°C)
15	Internally pressurize the bottles using compressed air at 0.68MPa
—	Depressurize and examine visually for signs of crazing
Repeat 25 times	To pass the test there should be no leakage from the pressurized bottles after the test has been repeated 25 times

7.5 Special tests for Returnable/Refillable PET bottles

Returnable/ Refillable (RR) PET bottles are containers that are designed to be re-used. They are typically washed in a hot caustic solution before rinsing and refilling and are therefore heat set to prevent them from shrinking.

A simple test has been developed to check that the degree of heat setting is sufficient to prevent unacceptable shrinkage during the hot wash: the bottle is submerged in hot water at 59 ± 1°C for 5 h. In order to pass the test, any capacity reduction should be less than 1%.

RR bottles tend to be thick-wall PET containers with a low degree of orientation in the shoulder and base regions and are therefore susceptible to environmental stress crazing (ESC) in these areas. A test has been developed to check the resistance of a bottle to ESC (Table 7.7).

7.6 Statistical Process Control

Statistical process control (SPC) is the application of statistical methods to the monitoring and control of a process to ensure that it operates at its full potential to produce conforming product.. Much of the power of SPC lies in the ability to examine a process and the sources of variation in that process using tools that give weight to objective analysis over subjective opinions and that allow the strength of each source to be determined numerically. Variations in the process that may affect the quality of the end product can be detected and corrected as early as possible, thus reducing waste as well as the likelihood that defective products will be shipped out to the customer. With its emphasis on early detection and prevention of problems, SPC has a distinct advantage over other quality methods, such as inspection, that apply resources to detecting and correcting problems after they have occurred.

The quantitative measurement of quality characteristics is the basis of SPC. The idea is to take enough samples to get to know the process being monitored; to find out whether the process is running well or not, and to learn the limits of the process. Every process has some variation which is the cumulative error of every element in the system. Sometimes there are *special causes* of variation in a process that can be identified and eliminated without too much difficulty. For example, when a heater band fails it can be replaced, or when temperature controllers are not properly calibrated they can be retuned. In other cases it may be more difficult to assign a cause, or quite possibly the expense of eliminating a known cause cannot be justified. However once the special causes have been identified and corrected, it becomes possible to determine the so called natural variation of the process by analysing data taken over long periods of time.

Thus there are two distinct phases involved in setting up a quality programme for a typical manufacturing process. *Phase One* is analytical. Quality control technicians evaluate the process to determine what data should be monitored and how often samples should be collected. Data is then collected to obtain a history of the process. A process description is derived from this history that identifies the centre of the process and the natural variation. If there is *special cause* variation in the operation, appropriate action is taken to eliminate it.

Phase Two of the programme involves using the process description as a control element for alarm monitoring and optimisation. Samples are collected on a regular basis and plotted on Control Charts in which the *warning* and *action* thresholds are derived from the historical data. Operators can monitor these charts to see if the process is stable and within limits.

SPC terminology

Feature/ Characteristic—the portion of the component or process under observation; usually the critical dimension(s) of the product.

N (sample size)—the sample size is the number of items in the sample.

Sample (subgroup)—group of items taken from the process either consecutively, or at pre-determined intervals

Variable—the characteristic of a part or component that can be measured such as a dimension in millimetres, the weight in grams, etc.

Attribute—a characteristic of a part or component that cannot easily be measured directly, such as whether a hole has been reamed, a serial number has been stamped, or that a certain operation has been completed. Also any measurements that are made in a YES, NO fashion, e.g. using a GO/ NO GO gauge.

Tolerance—a specification given to a size. A *unilateral tolerance* has only one value, e.g. the maximum permitted value, while a *Bilateral Tolerance* has both a minimum and a maximum value, e.g. a diameter of 30.00 mm ± 0.025.

Normal Frequency Distribution

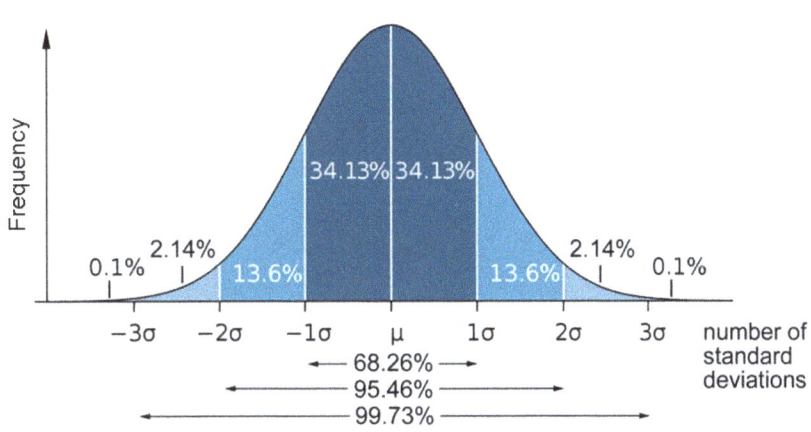

Percentages under the normal distribution curve

Control Charts

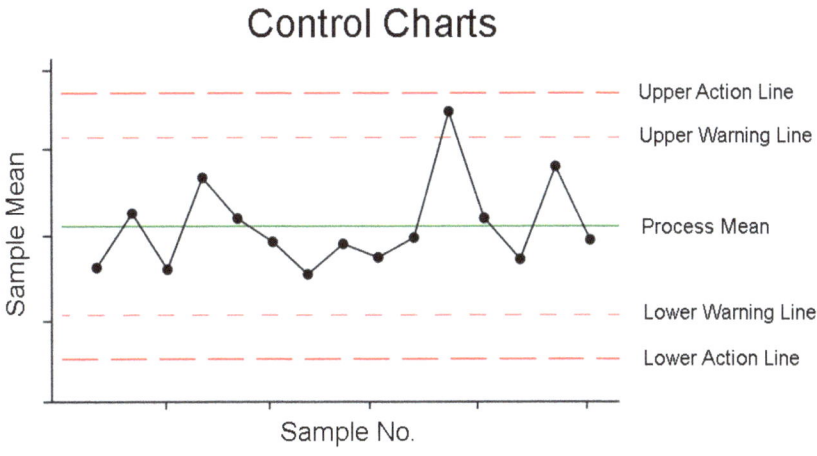

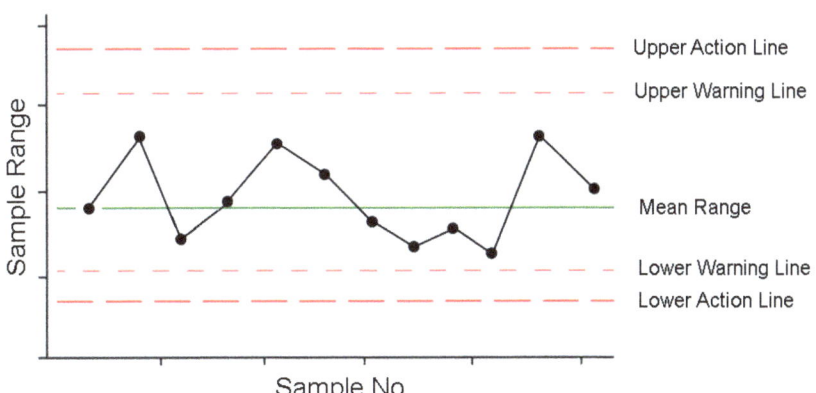

Control Charts—used to monitor, through Means and Ranges, a manufacturing process which is *in statistical control* (i.e. not changing its characteristics of average and spread). They visually highlight when action needs to be taken to correct a process that is no longer running satisfactorily. Chart limits are determined using statistically sound methods, and typically will include a centre line with *warning* and *action* lines on either side. The different zones are associated with defined rules for determining what action, if any, need be taken.

Histogram—the data is depicted in a vertical bar chart. Its shape reveals the frequency with which events occur, and the chance of any particular value being found. When analysing a frequency distribution there are three main points to monitor:

- the centre or average position.
- the spread or dispersion.
- the shape of the distribution.

With a *Normal distribution* the mean of the distribution is at the highest point, and there is an equal area under the curve either side of the mean.

Note that each coloured band in the normal distribution pictured on the previous page has a width of *one standard deviation*.

Standard Deviation—a measure of the spread or scatter of a population around its central tendency. Represented by the Greek letter sigma: σ

7.7 Continuous Process Control

A technique known as continuous process control involves building a mathematical *model* that embodies the relationship between key characteristics (e.g. part dimensions) and process parameters. Deciding which process parameters are best to use when building the model is not easy, but might include injection time, peak injection pressure, and screw recovery time.

Test runs during which critical process parameters are varied and finished part characteristics of the sample parts are measured, provide the data for the model. A good model will enable the computer system to recommend the optimum process settings, as well as suggesting process changes to keep a process on the straight and narrow. In an ideal world, such a system would do away with the need for inspection procedures altogether, although this won't be possible where the defects are aesthetic because such faults are often difficult to predict from process parameters.

7.8 Overall Equipment Effectiveness

In a model factory, equipment would operate 100% of the time at 100% capacity, with an output of 100% good quality. In real life, however, this situation is rare. The difference between the ideal and the actual situation is due to losses such as:

Downtime: refers to time when the machine ought to be running, but is not, either because of equipment failure, or the setup is being adjusted.

Sudden and unexpected *equipment failures*, are an obvious cause of loss, since a breakdown means that the machine won't be producing.

Most machine changeovers involve some period of shutdown while the tools are exchanged. This often includes substantial *time spent making adjustments* until the container quality is acceptable.

A *speed loss* means that the equipment may be running, but it is not cycling at its intended speed.

Minor stoppages are those caused not by technical failures, but by small problems such as parts that block sensors or get caught in machinery. Although the operator can easily correct such problems as they occur, the frequent halts can dramatically reduce the effectiveness of the equipment.

A *defect loss* means that the equipment is producing products that do not fully meet the specified quality characteristics. Defect losses include two major types of loss: scrap and start-up losses.

Scrap occurs when products do not meet quality specifications. The goal should be zero defects—to make the product right the first time, and every time.

Start-up losses occur when production is not immediately stable at equipment start-up, i.e. the first-off products don't meet specifications.

The diagram overleaf illustrates graphically how the losses in availability, performance, and quality come together to reduce the overall effectiveness of a machine.

The top bar *Total operating time*, shows the total time the machine is available for production. Bars A and B show availability. Bar A represents the *Net operating time*, which is the time available for production after subtracting planned downtime (no scheduled production) such as a holiday, no orders, or lack of personnel.

Bar B shows the *Actual running time* after subtracting downtime losses such as equipment failures and setup adjustments.

$$\text{Availability Rate} = \frac{\text{Operating Time} - \text{Downtime}}{\text{Total Operating Time}}$$

Bars C and D show performance. Bar C represents the Target Output of the machine during the running time, calculated at the intended cycle of the machine. Below it, a shorter fourth bar, D, represents the actual output, reflecting speed losses such as minor stoppages and reduced operating speed.

$$\text{Performance Rate} = \frac{\text{Total Output}}{\text{Potential Output at Rated Speed}}$$

Bars E and F show quality. As can be seen, the actual output (E) is reduced by defect losses such as scrap and start-up losses, shown as the yellow highlighted portion of bar F.

$$\text{Quality Rate} = \frac{\text{Good Output}}{\text{Total Output}}$$

As the diagram shows, the bottom-line good output is only a fraction of what it could be if losses in availability, performance, and quality were reduced. The diagram also suggests that to maximize effectiveness—to grow the good output on the bottom line—one must reduce not only quality losses, but also increase availability and minimize performance losses.

7.8.1 The benefits of OEE measurement

The aim of measuring OEE is to improve the effectiveness of the equipment. Since equipment effectiveness affects shop floor employees more immediately than any other group, it is appropriate for them to be involved in tracking OEE, and in planning and implementing strategies to increase effectiveness.

Ideally the machine operator will collect the daily data about the equipment for use in the OEE calculation. Collecting this data will:

- teach the operator about the equipment
- focus the operator's attention on the losses
- grow a feeling of ownership of the equipment

The shift leader or line manager is often the one who will receive the daily operating data from the operator and process it to develop information about the OEE. Working hands-on with the data will:

- give the leader/ manager basic facts and figures on the equipment
- help the leader/ manager give appropriate feedback to the operators and others involved in equipment improvement
- allow the leader to keep management informed about equipment status and improvement results.

		Total operating time		
Availability	A	Not operating time		No scheduled production
Availability	B	Actual running time	Failures & Setup	
Performance	C	Target output		
Performance	D	Actual output	Minor stoppages & slow cycle	
Quality	E	Actual output	Lost effectiveness	
Quality	F	Good output	Scrap & Start-up losses	

$$\text{OEE} = \text{B/A} \times \text{D/C} \times \text{F/E}$$

Availability rate x Performance rate x Quality rate

10

Process optimisation

Process variables are of two types: *controllable* and *consequential*. Controllable variables are the ones that the process technician selects to produce an acceptable container and include holding (packing) pressure and time, cooling time, barrel temperature, screw rotation speed, changeover position, shot weight, etc. Consequential process variables are the outcome of the interaction of the controllable values and include melt cushion, mould filling pressure and time (despite the presence of *injection pressure* and *injection time* "inputs"), screw recovery time, cycle time, the temperature of the preform when it leaves the mould, etc.

The aim should be to settle on those process settings that will produce consistent product of the required quality at the most effective cycle time, but how this is done will affect the outcome. For example if process settings have been set by different operators solely according to their own experience and understanding, the diversity of set-ups could be confusing for other operators. Also if process settings have been allowed to *evolve* as the result of multiple attempts to "process around" different issues, set-ups can be left on a knife edge (i.e. a point where intrinsic process variation is capable of dramatically affecting the quality of the product). To avoid these problems a standard process setting method that will lead to the optimal set-up in a logical and repeatable manner should be adopted.

As with any plastics process, the starting point should be the material and in particular *drying*. PET pellets, unlike those of PP, absorb water from the atmosphere during storage. This moisture does not harm the pellets, but when moist PET is melted, a chemical reaction between the hot polymer and the water known as *hydrolysis* causes the molecular chains to break, reducing the IV. To prevent this IV loss and the attendant reduction in physical properties, it is necessary to dry the PET using a dehumidifying system until the moisture level is less than 50 ppm by weight. Drying is covered in more detail in *Chapter 4—Drying*, but at its most basic involves subjecting the wet pellets to very dry conditions so that any absorbed moisture diffuses out. To accelerate this diffusion, the pellets are heated inside an insulated hopper to a temperature above 150°C. At this temperature the moisture diffuses from the PET pellets in as little as four hours. At lower temperatures, or when the PET is wetter than usual, longer drying times may be necessary. The pellets inside the hopper are heated by blowing hot air through them. The moisture-laden return-air then passes through a heat exchanger which transfers some of its heat into the entering air, before being channelled though a desiccant system that removes much of the moisture from the air, enabling it be re-circulated. (In older systems the heat exchanger was water-cooled and the heat simply

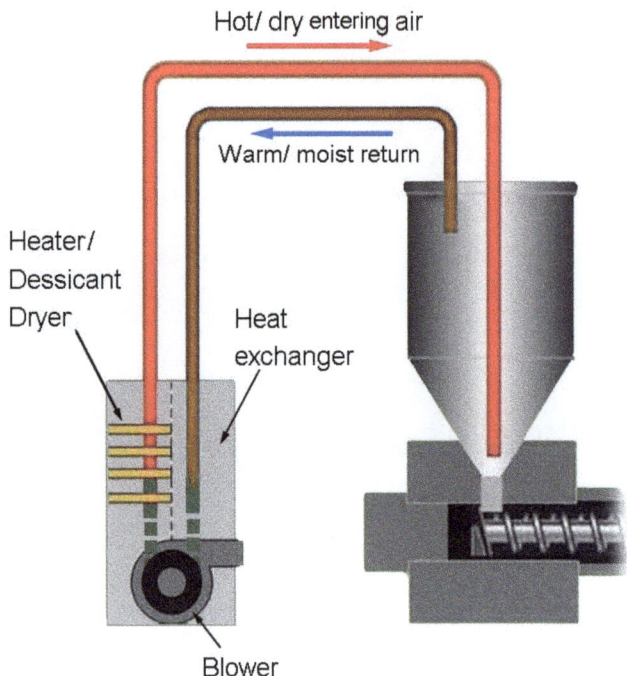

dumped into the cooling system. This meant that if the water flow through the heat exchanger was inadequate, the PET would not be properly dried, and so it was important to monitor the coolant flow rate in these systems). PET pellets contain PET dust "fines" which is picked up by the air flow and will obstruct flow through the desiccant if it is not first removed by a filtration system. (These filters need to be regularly checked to ensure that they have not become blocked and that the intended air volume continues to blow through the hopper).

Correct operation of the dehumidifying system can be confirmed by monitoring the *dew point* of the air entering the hopper. The dew point is the temperature at which water vapour present in the air will condense to form droplets of water. The lower the dew point temperature, the dryer the air. The dew point measured at the air inlet to the hopper should not be higher than -20°C. Ingress of ambient air must be prevented because dry PET reabsorbs moisture very rapidly. This means that if the PET is dried away from the machine, the dry pellets must be conveyed to the injection barrel throat in hot, dry air with a temperature and dew point similar to that of the drying air. If there is a prolonged interruption to moulding the dryer temperature should be turned down to 120°C to prevent the PET oxidizing (yellowing).

The material distribution in a container — which influences both its appearance and its physical properties — is largely determined by the heat profile of the preform from which it was blown (particularly so in the case of PP, which doesn't benefit from the self-healing stretching effect seen in PET). Fundamentally a preform's heat comes from the molten resin from which it was moulded, and thus the injection cylinder. When setting the injection cylinder heater band temperatures the first consideration should be how much of the *shot capacity* is to be used:

- If *about half* the capacity is used for each shot, the barrel temperatures should be set almost the same from back to front, or possibly slightly cooler at the feed end.

- If *less than half* the capacity is used for each shot, the barrel temperatures should be set significantly cooler at the feed end (to minimize degradation arising from the long residence time).

- If *more than half* the capacity is used for each shot, the barrel temperatures should be set almost the same from back to front or slightly higher at the feed end (to ensure the pellets are fully melted).

The melt temperature is not only governed by the barrel temperature controller settings, but also depends upon the temperature of the pellets entering the feed throat (which ideally will remain constant)[9]; the screw rotation speed (which should be set to give a surface speed of 250 - 350 mm/s — always provided screw charging is completed before the mould is ready to open); and the back pressure (which should normally be at the minimum setting).

Screw Diameter [mm]	Ø55	Ø60	Ø65	Ø70	Ø75	Ø80	Ø95
Minimum RPM	87	80	73	68	64	60	50
Maximum RPM	122	111	103	95	89	84	70

Recommended RPM for each screw diameter based upon an optimum screw surface speed for PET of 250 - 350 mm/s

Different screws add different amounts of shear heat, but it is common to see melt temperatures 10-20° C above the injection cylinder settings. A clue that the melt may be hotter than necessary is if the hydraulic pressure required for screw rotation is unusually low (ideally the hydraulic pressure "used" by the machine will be around *half* of what it is capable of providing).

At the instant when charging of the cylinder has just been completed, the heat distribution in the melt is likely to be patchy, and so for PP in particular it can be advantageous to set a short delay between the completion of charging and the start of injection, in order to provide sufficient time for the heat to become more evenly distributed.

The temperature of the melt influences its viscosity and density as well as its degradation rate (and with PET the AA level). Keeping the temperature low ensures the density of the melt is high and thus reduces shrinkage in the mould.

---

[9] Measuring the throat temperature and establishing a low temperature alarm point is a straightforward way of monitoring the stability of the material drying process.

Colours represent melt temperature variation

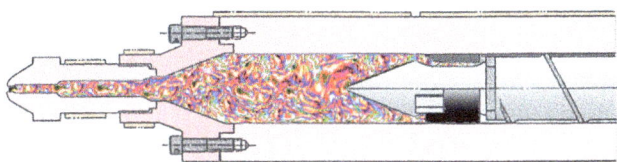

The instant when charging of the cylinder has just been completed

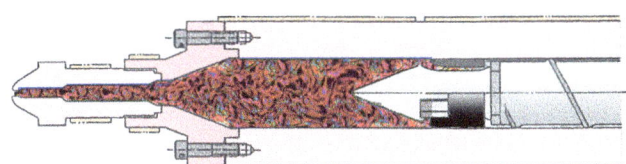

A short delay between the completion of charging, and the start of injection can be advantageous to provide time for the heat to become more evenly distributed

The **Hot Runner System** should be set at the temperature of the *polymer*, so that it maintains but doesn't alter the melt temperature. Keep in mind that frictional *shear heating* will have raised the melt temperature, so the hot runner temperature setting can be higher than that of the injection barrel nozzle (but typically by not more than 5°C). It is always wise to check the *gate balance* by pressing the emergency stop part-way through injection to produce a set of *short shots*. Any minor unbalance can then be corrected by raising or lowering the temperatures of the individual HR nozzles.

The **shot size** not only determines the amount of plastic that will be injected into the mould, but also the *cushion*, which is the amount of screw travel remaining after the screw has reached its most forward point at the end of the holding phase. Without this cushion it would be impossible to transmit hydraulic pressure into the mould and prevent sink marks; although too large a cushion means longer residence times leading to degradation and increased acetaldehyde (AA). First set the shot weight to be too large, and then gradually reduce it until the correct cushion is achieved. As a *rule of thumb* the cushion left at the end should not be more than one tenth of the screw diameter.

Injection involves two different pressure settings: the **injection** pressure setting (V81) limits the maximum pressure available when the melt is being injected into the cavities, while the lower **holding** or "packing" pressure setting (V84) is applied to prevent the mould being flashed and to squeeze additional material into the cavity to compensate for the volume lost as the polymer melt cools and contracts. Picture the injection screw and barrel acting like a hypodermic syringe: hydraulic oil pressure is applied against the injection ram to push the screw forward like a piston and inject the melt into the cavities. The injection pressure needs to be high enough to inject the material through the injection nozzle and hot runner system and into the cavity against the resistance stemming from the viscosity of the melt, as well as the resistance arising from compressing the material. However when the cavities are full and there is no free space remaining to accommodate any more melt, if the pressure continues to be applied, the pressure in the cavity will rise abruptly. To prevent this the available pressure is reduced at the point known as the "velocity to pressure switchover" (or **P/V point**), being the position in the screw travel where the injection screw movement is changed from *velocity* control (applied during injection when the screw is moving) to *pressure* control (applied during holding when the screw is no longer moving). The position where the cavities become full of melt can be identified approximately by the sharp rise in the injection pressure at this point. The change-over is triggered either by a proximity sensor actuated by the screw, or on suitably equipped machines, by the injection controller. If the change-over is set too far *forward*, **flash** (attempting to overfill the cavity using high pressure) can occur. If it is set too far *back*, **under-packed** preforms, or in extreme cases, **short shots** may be the result.

Watch the pressure gauge when injecting manually. Injection is over so quickly it can be difficult to register the peak pressure. On machines without an injection controller, assuming the appropriate sensor has been triggered, the holding pressure (V84) will be displayed for as long as the switch is held in the "inject" position, making it easy to read. In order to verify the set injection pressure, manually inject a set of preforms, then momentarily switch to "neutral" and back to "manual" (this resets the changeover signal from the proximity sensor), and operate the "inject" switch once again to display the injection pressure (V81). This time there can be no change-over signal from the proximity sensor because the screw is already past this point, and so the (available) injection pressure is displayed.

The injection pressure is limited by a hydraulic Relief Valve (V81) which should always be set a little under the maximum pressure permitted by the machine design (typically 13.7 MPa), and the injection speed control (V75) should be used as its name suggests, to set the screw velocity. Always keep in mind that whatever injection pressure is set, the machine will only utilise sufficient pressure to deliver the set injection speed. Should this happen to be fairly slow, then the machine will not need much pressure to achieve it.

In other words the hydraulic pressure in the system automatically regulates itself to any necessary pressure *up to the maximum available* on V81, to achieve the specified velocity. (The benefit of which being that if the viscosity of the melt should happen to fluctuate slightly during moulding, only the hydraulic pressure will vary, leaving the injection speed — and thus filling time — unchanged).

The proper **injection speed** varies from bottle to bottle and depends on preform length and wall thickness. Generally a reasonably fast injection speed should be used, since otherwise the preform will have a steep temperature gradient from gate to finish that can affect the material distribution in the container. With a fast injection, each part of the preform is cooled by a similar amount, but with a slow injection, the last part of the preform to fill — the interior of the tip — has less time for cooling and so remains hotter, meaning this area will stretch more easily during the blowing process. (Although injecting much too slowly can allow the gate to commence freezing off before injection is completed, resulting in crystallinity around the gate).

Fast injection: very uniform temperature top to bottom

Slow injection: extreme temperature differential top to bottom

On the other hand the injection must not be so fast that shear heating leads to streaks and degraded material in the preform. While a fast injection speed does increase shear heating, the situation is complicated by the fact that melt viscosity also alters according to the flow rate. As the flow speed is increased, a phenomenon known as sheer-induced orientation means the molecular chains line up in the direction of flow, thus enabling them to move more easily, and so shear heating actually decreases.

Injection time (T00) should be long enough to fill the mould *and* pack it out sufficiently to eliminate sinks, voids, and waves. The melt starts to cool as soon as it enters the cavity which means that the injection time also serves as cooling time. (In fact the cooling effect is greatest during the holding phase due to the higher pressure).

The packing pressure also influences how thermal gates close off. Insufficient packing will not cool the material in the gate adequately, resulting in stringing, while too much pressure (over-packing) can mean plastic re-extrudes back out of the preform into the hot runner nozzle once the pressure is released. Since this plastic has already begun to cool in the mould and start to crystallize, when it is carried into the cavity in the next shot, it appears in the preform as a white slug of crystalline PET near the gate.

As a *rule of thumb* set the **hold pressure** to be between 3.0-4.5 MPa for wide mouth, or 5.0-7.5 Mpa for narrow neck containers. This is a good starting point for a new process and can be fine tuned later. Machines with injection controllers have a second hold pressure that is usually unnecessary, being merely an extension of packing pressure at a different pressure setting. This setting can be set the same as the packing pressure unless experimentation yields better results.

Generally speaking, if any of the following are *increased*, the preform will be cooler:

- injection time — because there will be more time in the cavity for packing and thus cooling
- cooling (curing) time — because the preform will be in the cavity for longer
- injection speed — since the cavity will fill faster, reducing top to bottom temperature differences, and leaving more time for cooling during packing
- hold pressure — the effect of forcing the plastic more firmly against the mould will be to improve cooling (but only until the gate closes).

Rule of thumb:

T00 + T01 [secs]

= (preform wall thickness [mm])² + 1

To determine the gate freeze/ close time, having first ensured the machine is moulding consistently and that there is an adequate cushion of melt remaining in front of the screw at the end of the packing phase, make a succession of preforms, each time reducing the *Injection time* and increasing the *Cooling time*, (so that the combined total remains the same). Weigh the preforms and tabulate the results. The time when the preform stops increasing in weight is the point when the gate has frozen, or — in the case of a valve gate nozzle — can be closed. In the example below the preforms start getting lighter when the *Injection Time* is less than 12 seconds — indicating that the holding pressure is now stopping too early, i.e. while the gate is still open and before the cavity is fully packed, indicating that the correct *Injection time* setting would be 12 seconds.

Injection time T000	Cooling time T001	Preform weight [g]
12.50	5.50	70.0
12.25	5.75	70.0
12.00	**6.00**	**70.0**
11.75	6.25	69.8
11.50	6.50	69.5

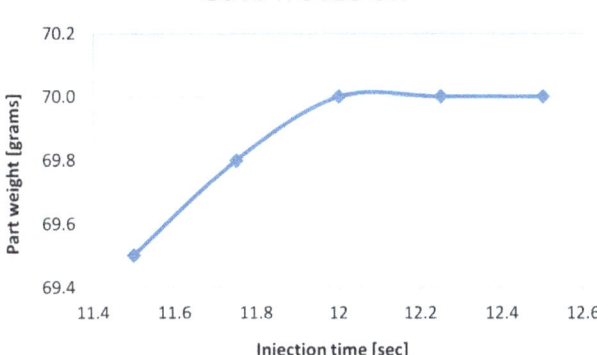

The injection mould temperature affects the **cooling time** (T01) and therefore the overall cycle time, as well as the preform temperature and thus the material distribution in the container. Since the correct cooling time can only be decided once bottle blowing has begun, at the start it is necessary to use an approximate time that will allow bottles to be blown. As *a rule of thumb*, the injection and cooling times [secs] when added together should equal the preform wall thickness squared plus 1 (where the preform wall thickness is in mm). A simple way to judge whether PET preforms are at a suitable temperature for blowing is to take two which are still hot and press them together. If they stick to one another, then they are capable of being blown.

The **direction**, the **temperature**, and the **flow rate** of the **chilled water** circulating through the injection core pins and cavities are important. The hottest part of the cavity is at the gate and so the coolant flow should be directed to this region first. A large temperature difference between inlet and outlet can be a warning that higher flow rates are required to remove the heat. (The optimum condition for heat dissipation and removal being to have no more than 1 - 2°C difference in temperature from inlet to outlet). As a simple *rule of thumb*, and assuming that the water contains corrosion inhibitors and ethylene glycol (antifreeze), at least 0.3 times the cooling channel internal diameter (mm) is needed in litres per minute flow rate to achieve turbulent flow. Typical mould temperatures range from as high as 170°C when moulding PES (when the fluid would need to be oil), to as low as 10°C when moulding PET. Water is used for the cooling medium when the mould temperature is below 90°C. Faster cycle times may be possible with colder chilled water, although condensation on the injection core pins must be avoided, and gates freezing can become a problem during production stoppages. To avoid gates freezing, leave the chilled water turned off until after the first manual shot has been injected (and remember to turn off the flow of chilled water if moulding is halted, even for a few minutes)!

A small amount of **back pressure** (creating resistance to the backward travel of the screw during charging/ plasticisation) increases the melt density of the polymer, but also results in increased shearing — and therefore heating — of the material, potentially leading to sink marks in the preforms and increased AA in PET bottles. Most one-stage applications do not require much back pressure to create a good process, and so it is normally set close to zero either by Relief Valve V84, or by means of the injection controller. However in certain instances such as finding bubbles in the melt, or when a coloured masterbatch is being used that needs extra mixing to improve its dispersion, increasing back pressure can be helpful. Typically back pressure is 0.7 to 2.1 MPa (100 - 300 psi).

Minimize drooling (on machines with retracting Injection Units) by setting the timers so that the Shut-Off nozzle (T43) remains open for 0.5 seconds after Injection (T00) finishes, and so that the Injection Unit remains pressurized forward (T44) for 0.10 seconds longer than the Shut-Off Nozzle is

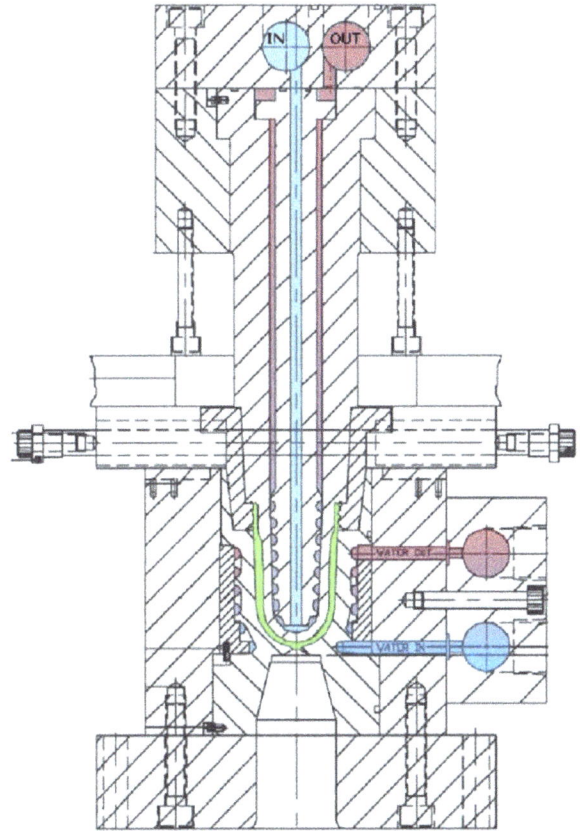

Cross-section though a generic one-stage injection cavity showing the coolant flow path

open. To further reduce drooling on machines fitted with an Injection Controller, use the second of the two holding pressure steps to lower the pressure in the Hot Runner Manifold before the Injection Unit retracts, by setting it to 10 kgf/cm².

"Stringing" is related to the hot runner nozzle temperature and (in valve gated systems) the temperature of the shut-off face of the pin. If the temperature is wrong, as the preform is extracted from the injection cavity a "string" like the one shown in the above picture, can be drawn from the sprue remnant. The correct temperature setting may be either hotter *or* colder. A hotter nozzle will mean that the melt has too little strength to draw a string. A colder nozzle means the melt is brittle, and therefore breaks instead of drawing a string. To find out which is best, try both hotter, and colder nozzle temperatures in turn, and select the temperature setting that cures the stringing problem without introducing any new problems such as gate imbalance or white crystalline preform tips.

The other main reasons for stringing to occur are:

(i) the NOMEX insulation paper surmounting the hot runner nozzle is damaged. Replace it.

(ii) the holding (packing) pressure is too high. Reduce it and also extend the hold time.

(iii) the shut-off pin is not closing correctly. Check the hydraulic actuator function.

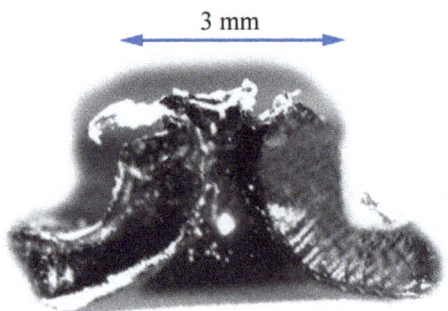

"Pinhole" through gate

An occasional problem with thermal gates is the appearance of a shrinkage void or "pinhole" in the sprue or gate remnant. Longer hold times and lower hold pressures will feed material to the sprue at a rate that will eliminate voids without over-packing it. Overall cycle time does not have to be extended if the cooling time is decreased by the amount the hold time is raised. Avoid setting high levels of manifold decompression.

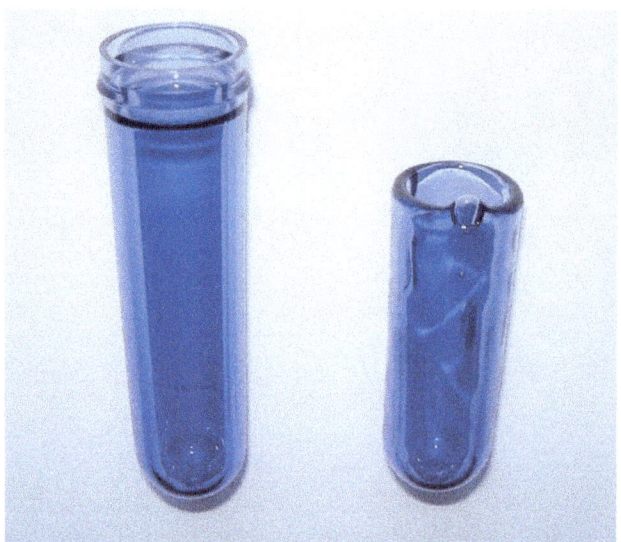

The short-shot on the right "freezes" that instant in time when an air bubble emerges from the advancing melt front.

Eliminating bubbles & splay marks

Bubbles can be formed in preforms for a variety of reasons, the most likely being inadequate drying of the PET. Another possiblity is the "sprue-break" movement necessary on Nissei ASB-50, ASB-250, and ASB-650 machines where the injection nozzle must draw back from the hot runner manifold inlet before the lower mould can be opened. (In the case of thermal-gate nozzles, the "sprue-break" movement has an important secondary function of decompressing the hot runner manifold before the preform is withdrawn from the mould. Otherwise the molten polymer inside could well up into the cavity and freeze-off, thereby blocking the gate).

Sometimes the forward movement of the injection unit (i.e. after the mould has closed ready for the next cycle) can trap air in the manifold inlet which is then injected into the cavity together with the molten polymer as a bubble (see picture above).

"Splay marks" are generally the result of an air bubble emerging from the advancing melt front after the PET has begun to solidify, although they can also be caused when cold material that has drooled from the manifold inlet is pushed back into the inlet by the forward movement of the injection unit and subsequently injected into the preform where it can leave a witness mark. Although bubbles can and do occur, splay marks seldom occur on machines that do not use "sprue-break" such as the Nissei ASB-70DPH or Aoki SBIII machines.

Hot runners are designed to ensure that there is a sufficient volume of melt passing through the system to flush out any entrapped air so that any bubbles

"Splay marks" are visible in this PET bottle wall.

emerge from the advancing melt front *before* the PET begins to solidify; but if bubbles or splay marks do occur, they can usually be eliminated by applying the following guidelines:

a) Reducing the holding pressure will mean that less melt "drools" from the inlet when the mould is open (it is the residual pressure inside the manifold which forces the "drool" from the inlet), so there is less opportunity for air to be entrapped.

b) Increasing the temperature (by increasing the back pressure) or shearing the material more (by increasing the screw RPM), will make the melt hotter and therefore less viscous, which will speed up the travel of any bubbles.

c) Reducing the hot runner sprue bush or *inlet* temperature, so that the viscosity of the melt is increased, will decrease the tendency of the PET to drip from the inlet.

d) Ensuring that the timers controlling the process are set correctly — in the diagram below the time *sequence* is represented by the relative position of the different arrows, and time *duration* by the length of the arrows.

Blow Station

The main adjustments at the blow station are Primary and Secondary blow delay, primary blow air pressure and flow rate, and Stretch Rod speed.

Bottle **blowing** takes place in two steps: first Primary air, together with the Stretch Rod, *places the material in the desired position;* and then the Secondary air presses the material firmly against the blow mould surface for good shape definition and rapid cooling. The Primary and Secondary blows happen in very rapid sequence and often the effects of the first are *masked* by the latter. Therefore when starting to blow a bottle for the first time, initially use only Primary air pressure.

Stretch speed is a very important factor in achieving good molecular orientation during blowing. For PET, the faster it can be stretched, the greater the orientation. A key to attaining good stretch rod speed is to make sure the machine is equipped with "quick exhaust" on the stretch rod cylinder. This consists of a valve mounted in the cylinder exhaust port that will allow the exiting air to vent immediately when stretching. Also, an adequate machine system pressure between 0.96 and 0.98MPa will assure strong stretching. Orientation is not the only quality that is affected by stretch speed. It can also influence top to bottom material distribution of the container (although this is not the usual method of adjustment). In general, a faster speed will help pull material out of the neck, while a slower speed will do the opposite. Remember that blow delays will need to be adjusted accordingly when experimenting with speed.

T00[10] Injection time (filling and holding)

T02 Delay before retraction of injection nozzle

T03 Delay before screw charging starts T41 Delay before injection starts

T43 Shut-off nozzle remains open for this time after injection has finished

T44 Injection nozzle pressurized forward until this times out

---

[10] Fuji Controller as used on Nissei ASB machines

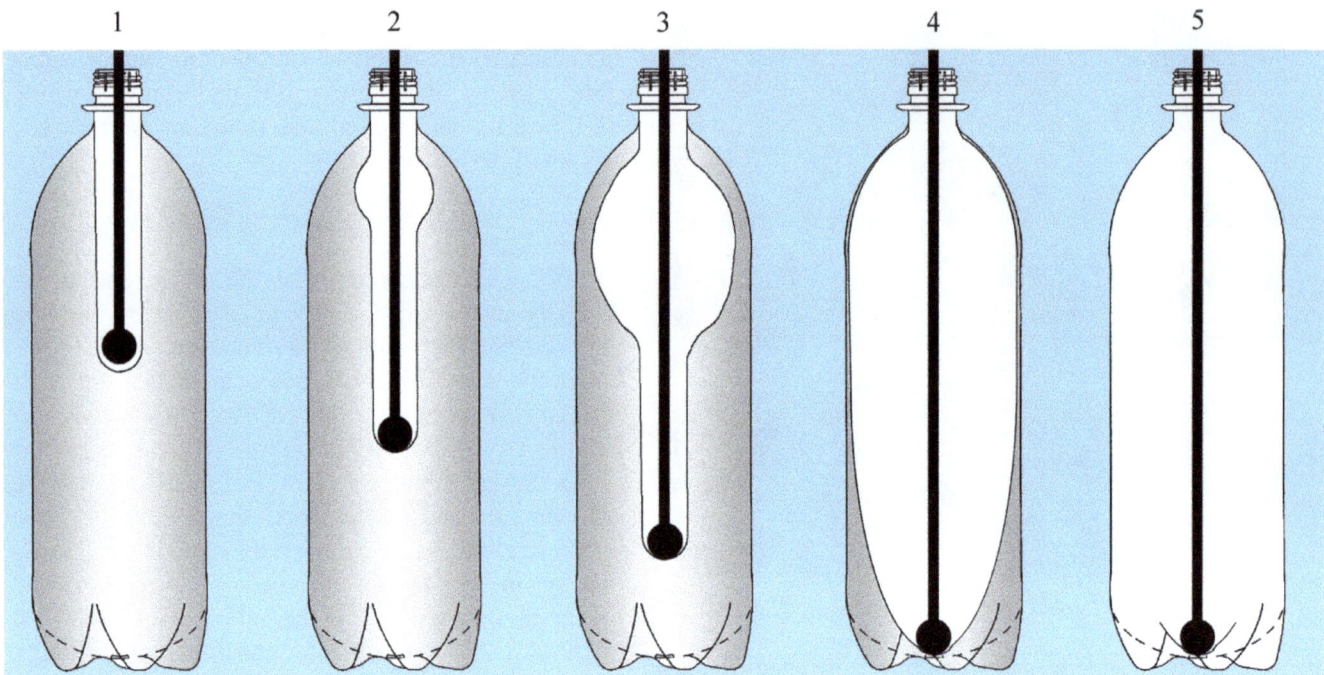

1. Hot preform in blow mould, stretch rod inserted into preform
2. Stretching underway and primary blow has just started
3. Stretching and blowing continue
4. Stretch Rod at maximum travel, material distribution in bottle now largely determined
5. Secondary blow presses bottle against blow mould for maximum definition and cooling effect

Primary blow air pressure has a major influence on the material distribution in the container, since it controls how the preform expands in the initial stages of stretch-blowing. A low blowing pressure will allow the stretch rod to extend further before the preform starts to expand, thereby carrying more material towards the base of the container; although a pressure that is too low can mean the preform touching the rod resulting in a cold (thick) spot in the container wall (or *loupe* from the French word for lens), and in extreme cases the tip of the preform can envelop the stretch rod tip creating a *crater* (massive build-up of material in the bottom of the container accompanied by folds of plastic). With a higher blowing pressure the preform will expand more quickly and tend to freeze the plastic off in the uppermost part of the container. If the blowing pressure is far too high, the preform expands so quickly that the gate is blown off-centre.

Initially all primary air flow control valves should be fully open, and conditioning pots and cores should be operating for a *minimum* time. A good starting point for low air pressure is 1.0MPa. Then adjust the Primary air pressure, Primary air blow timing, and Stretch Rod speed to give good material distribution in the bottles.

The primary blow delay time can also have a significant influence on the thickness of the shoulder and the base of the bottle. Increasing the pressure will mean the preform inflates more rapidly, leading to a heavier shoulder/ lighter base. Delaying the primary blow will have the opposite effect, leading to a thinner shoulder/ heavier base. It is possible that the two process changes could even cancel each other out.

The presence of **pearlescence** in a PET container indicates that part of the preform was too cold (or at least not hot enough) to stretch the required distance. (Pearlescence has the appearance of white *mother of pearl* as its name suggests when examined under *reflected* light, but when looking *through* a pearlescent section at the sky (i.e. *transmitted* light), it appears brown).

Secondary blow air pressure shapes the final container, dictating the surface definition and also influencing the capacity of the container. It not only affects how rapidly the preform inflates, but similar to the stretch speed, the level of orientation induced in the PET molecules. In general, just sufficient air pressure should be used to create a well-formed container; too much can force the moulds apart resulting in wide *seams* on the final product. Air pressure must be maintained in the closed mould long enough for the PET to cool and become rigid. If the blow time were to be too short, the container would likely have flats at the blow mould parting lines where the material had remained hot and sunk.

Some machines are equipped with **flow controls** for each primary blow air supply pipe. Their function is to individually control air flow to each cavity to enable fine tuning of the blow process. Normally, if all the preforms are correct, these should all be set fully open. Adjustments to these can be equated to adjusting primary air pressure to each cavity and has much the same effect.

Primary blow delay is the time before the primary air valve(s) open. This timer usually starts when the stretch rods are triggered to come down. This delay has much the same effect as adjusting the primary blow air pressure, but with the added benefit of being able to control when the preform starts to expand in relation to how far the stretch rod has extended. This can be very useful for adjusting material distribution up or down the container. A short delay will keep material in the top of the container, while a long delay will let it stretch to the bottom. The limiting factors are *off-centre gates* (when the delay is too short), or *craters* (when the delay is too long).

If no noticeable change in the material distribution is seen when altering the primary blow pressure or timing, *increase* the high pressure blow delay time (even to the extent that it never happens), in case the secondary blow is coming on too soon, and therefore not giving the primary blow air enough time to do its job.

Secondary blow delay timing starts at the same time as the primary timer starts. While the Primary air controls the material distribution in the container, the secondary air completes the process by using high pressure to ensure good definition and the proper container volume. There are exceptions to these guidelines. One is when a preform is not stretched very far (i.e. the axial blow ratio is very low). In this situation it is sometimes better to stretch-blow very quickly. This will increase orientation and so help material distribution, and since the amount of stretching is low will not result in off-centre gates.

The Blow air should be exhausted and the Blow Moulds open before the Injection Mould opens, otherwise the machine cycle time will be lengthened.

On most machines it is possible to delay the point

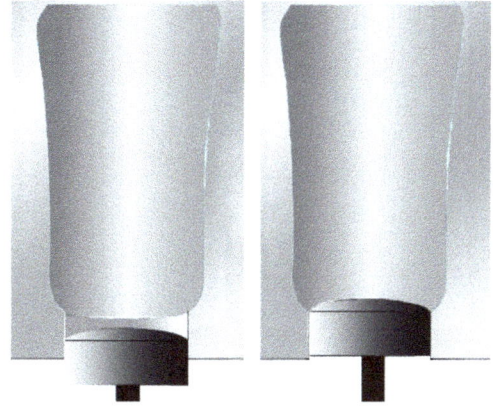

when the **Bottom Mould** rises during the blow moulding cycle. This can have considerable benefits for containers with a concave bottom (as in a champagne bottle), because when the bottom mould is in its DOWN position the preform can be *overstretched* in order to increase the amount of material dragged into the bottom of the container (see image above). Most **mould movement** timers will not affect the process, but the timing of the bottom mould UP movement is critical for the proper effect. If done correctly it can enormously improve the thickness and/ or definition of the container base.

On four-station machines, should bottles be non-uniform, **conditioning** can be used to individually alter preform temperatures to equalize any variation persisting from the injection moulding process. It can also be used to *profile* the preform temperature (i.e. warm any region of the preform that is required to stretch more easily in order to form a thin section in the bottle). Remember to check timer settings to confirm that the conditioning tooling is operating for long enough to have any effect on the preform.

Once the process has been optimised, ensure the process settings are recorded (and remember if changes are subsequently made, update the record).

Timing Charts

Aoki SBIII-250LL-50S machine Timing Chart. The Timers, Sensors, and Hydraulic Valves involved in the "injection" process are shown in sequence:

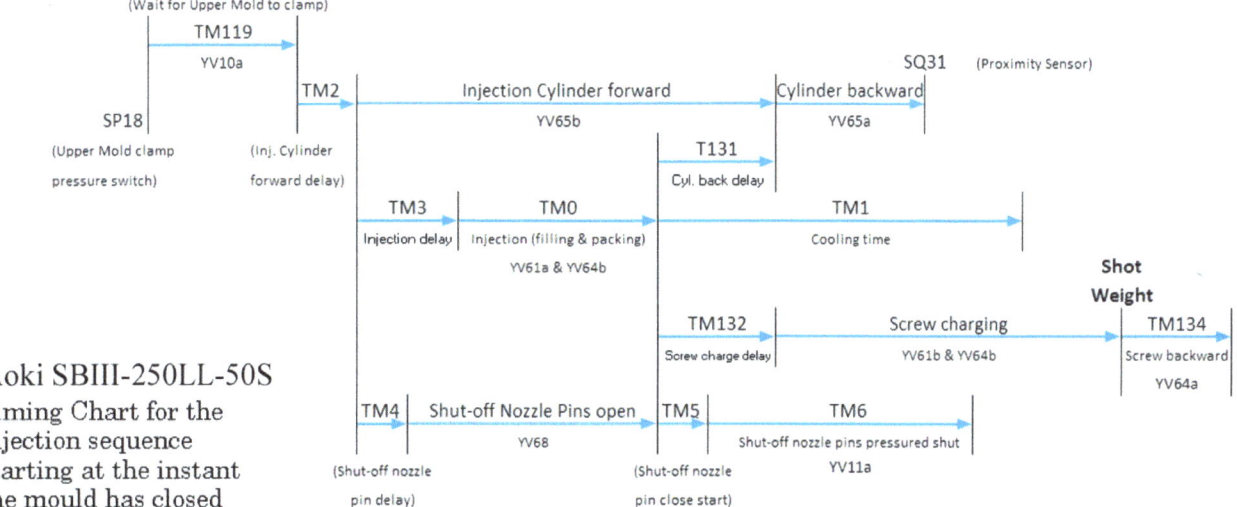

Aoki SBIII-250LL-50S Timing Chart for the injection sequence starting at the instant the mould has closed

Chapter 10—Process optimisation

Aoki SBIII-250LL-50S machine Timing Chart

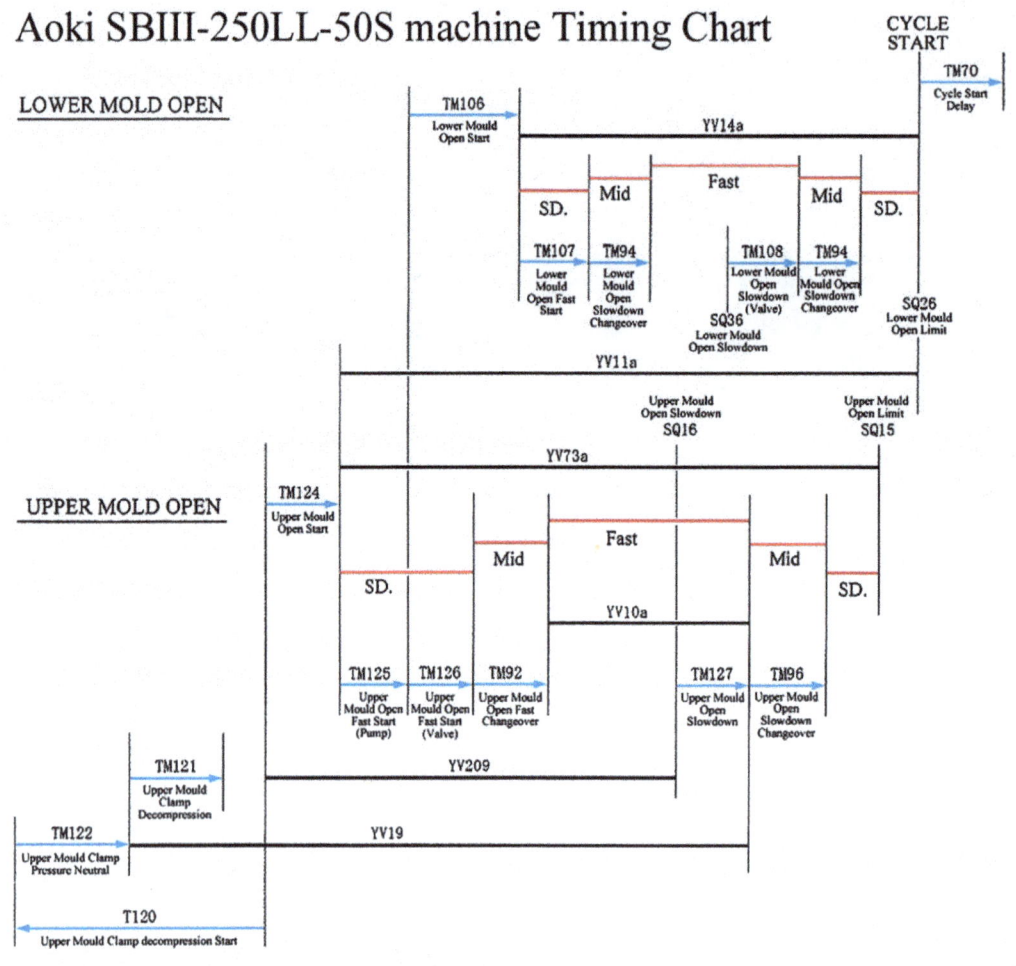

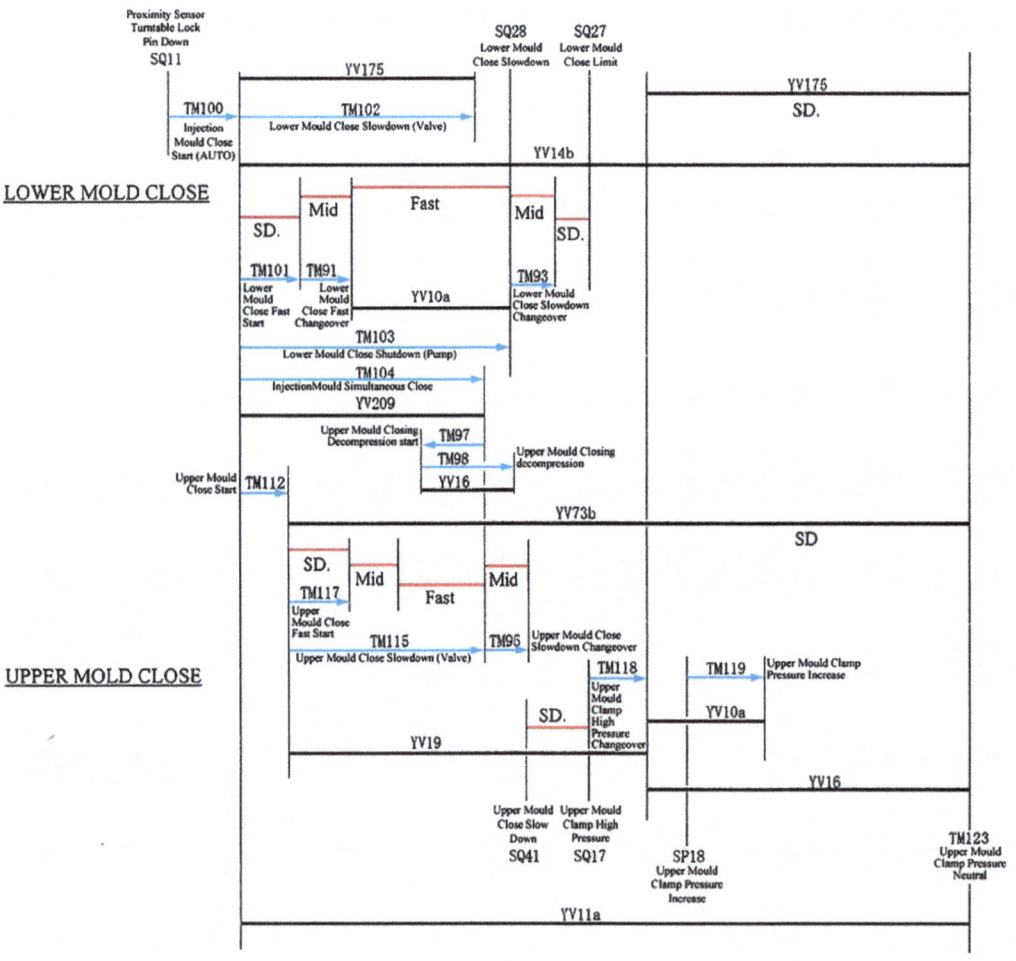

Chapter 10—Process optimisation 111

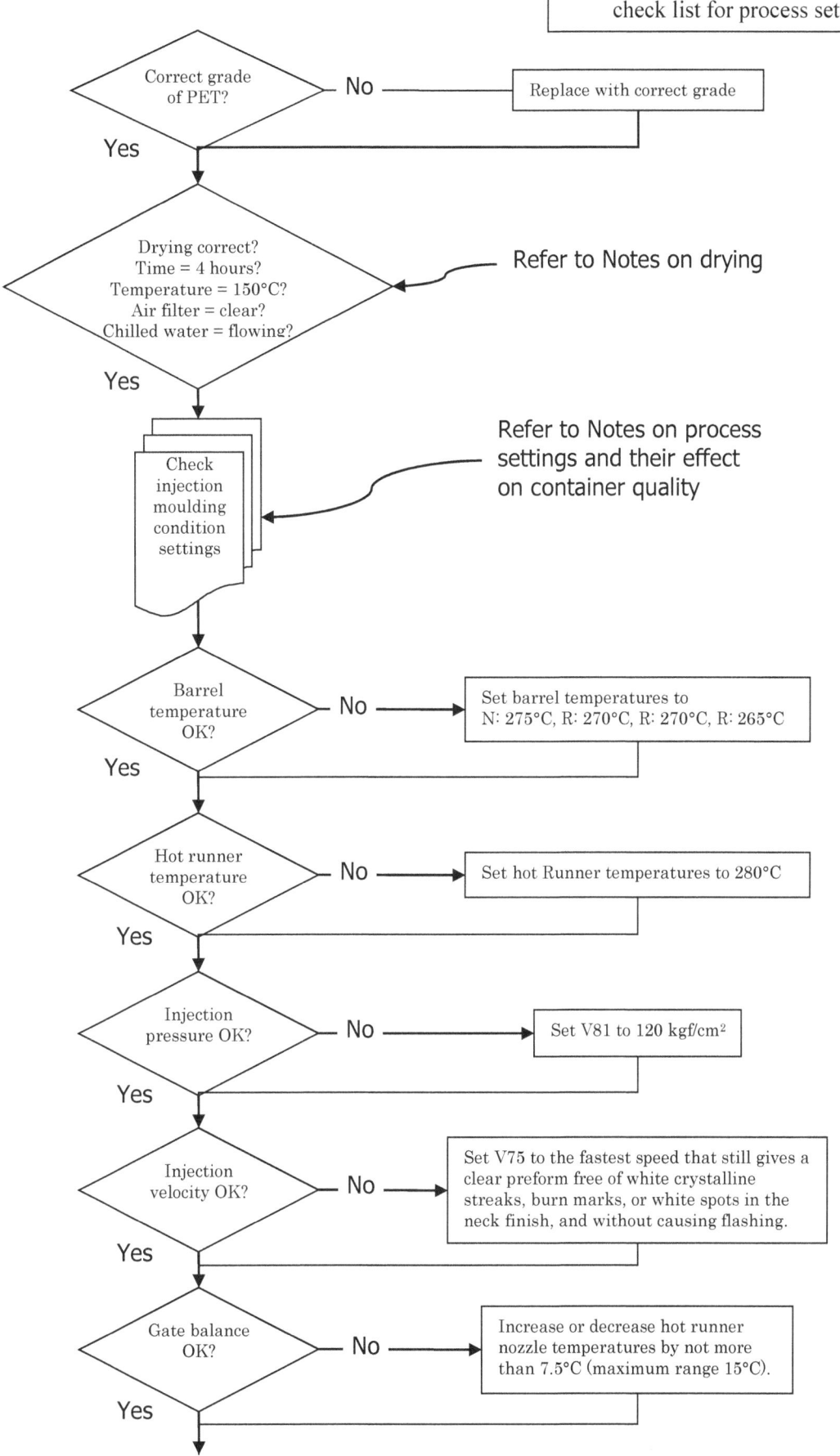

Chapter 10—Process optimisation

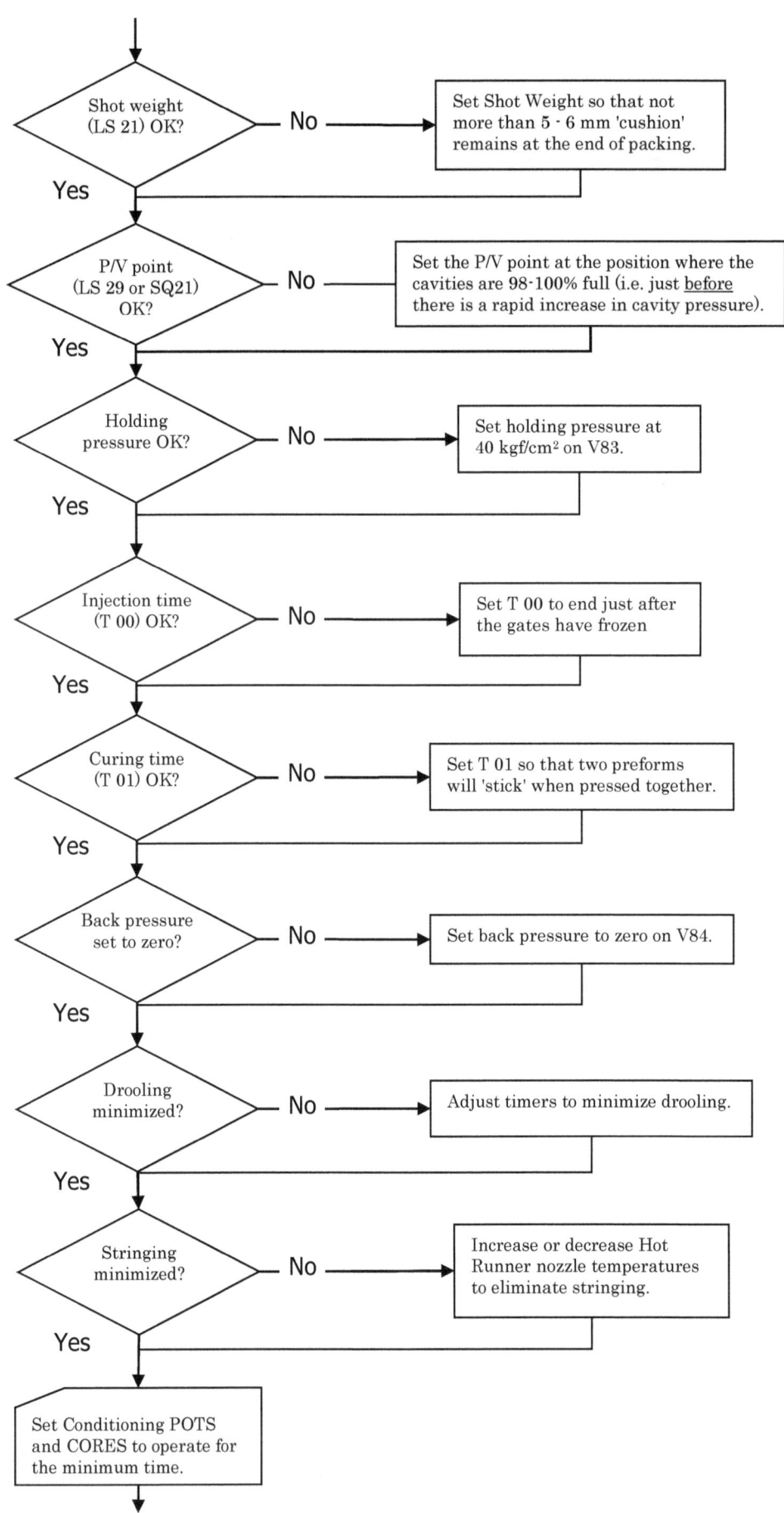

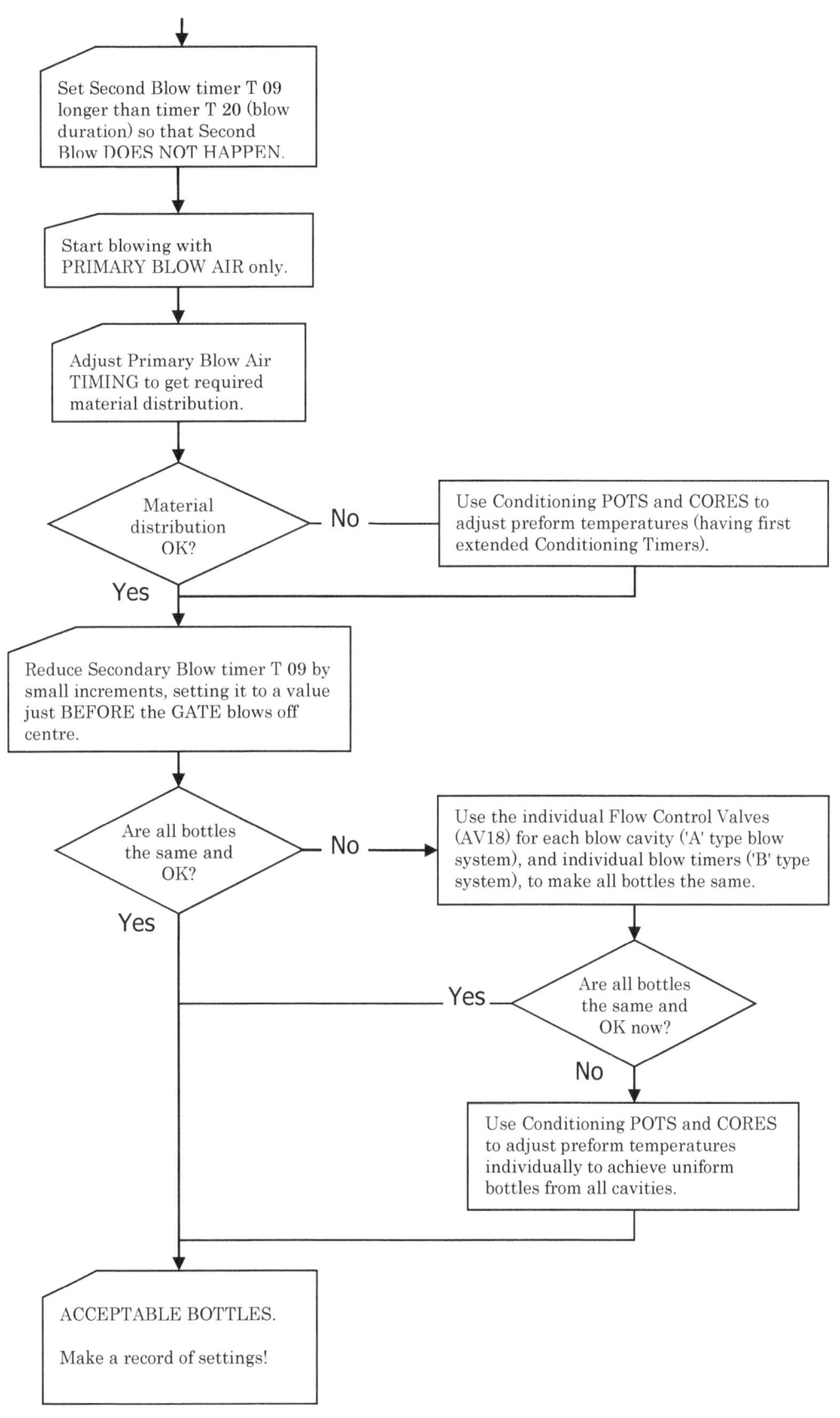

11

Troubleshooting

Troubleshooting - "the act or process of locating the cause of a fault and rectifying it." Successful troubleshooting requires:

- Adequate knowledge of the principles of operation to understand how a system should work (without this troubleshooting is nearly impossible).
- An ability to analyze problems in a logical manner.
- Proficiency in using any test equipment that may be used and in interpreting their readings, also in interpreting system schematics and parts lists.
- Skill in operating the machine.
- An understanding of how each process change affects the preforms and/or containers.

This manual provides a basic introduction to troubleshooting, but the required skills must be developed through experience with the machine and the process.

> *Knowing is not understanding. There is a great difference between knowing and understanding: you can know a lot about something and not really understand it.*
>
> Charles Kettering

Most problems fall into one of two categories: being either repeatable or intermittent. A repeatable problem is one where the problem occurs in response to a particular event (or sequence of events). If a problem is repeatable, this at least offers some initial clues about how to find the cause; and where alternative solutions are being evaluated, also provides a means of testing to see whether the problem has been resolved.

Intermittent problems are much more difficult to deal with as they appear to be random events without obvious cause and cannot be repeated. They can be extremely difficult and frustrating to diagnose. If possible spend time determining what the circumstances are when the problem arises. Sometimes problems that seem intermittent really aren't, it's just that the specific circumstances that trigger the problem may be ambiguous or not easy to recognise. Since truly intermittent problems cannot be duplicated at will, it may not be feasible to systematically work one's way towards the ultimate cause; making it necessary instead to employ trial-and-error methods: in other words making a change and then waiting to see if the problem recurs.

Sometimes problems involve more than one component or subsystem. The difficulty is figuring out which component is responsible for the problem. By making small logical changes, or replacing a component and observing the outcome, it is usually possible to narrow down the problem using a process of elimination. The key is to make only one change at a time and then see if the problem is resolved; if it is, then whatever has been changed is likely to be responsible for the problem (although in some cases it could be curing the problem indirectly).

If more than one process change is made at a time, it may be impossible to know which one was responsible for fixing the problem. If multiple changes have been made together, try undoing them one at a time in order to identify the critical change.

Problem solving techniques

A step-by-step approach helps ensure success:

1. define the problem
2. gather data to illustrate the problem
3. determine possible causes
4. select the root cause
5. devise solution strategies
6. test and evaluate solutions

The goal should not be simply to correct the effects of the problem, but to determine the root cause (so that it can be prevented from happening again in the future). One simple technique to do this involves repeatedly asking the question *why?* (five times is a good rule of thumb). Very often the ostensible reason for a problem can actually lead to another question, so that by the time *why?* has been asked for the fifth time, one can hope to have uncovered the root cause. It isn't always that simple, but the exercise can be helpful in figuring out what is really going on, and forestall leaping to "quick fix" solutions that don't resolve the root problem. It is especially useful for tackling chronic problems that show up again and again in a system; it is less useful for problems that are unlikely to recur. Here's an example:

Why? does one container have a different wall thickness profile from the others, despite a set of partial short shots showing all the cavities filling at the same rate.

The preform weight is correct and its wall thickness is uniform, so *why?* does this particular preform not have the same heat profile as the others.

Why?, when the respective injection core pin is exchanged with its opposite number, does the fault 'move' with the core pin (implying it is not having the same cooling effect as the other core pins).

Why? is the "bubbler tube" inside this core partially blocked with sludge.

Why? is the automatic water treatment system not doing its job and preventing the build up of sludge in the system.

(The method works best with on-the-spot verification of the answer to the current *why?* question before proceeding to the next).

Brainstorming with colleagues can assist the listing of possible causes. It is common for brainstorming sessions to turn up dozens of factors, and so a number of techniques have been devised to group the factors such as Affinity Charts[11], Mind Maps™, and Cause & Effect diagrams (colloquially known as Fishbone Diagrams).

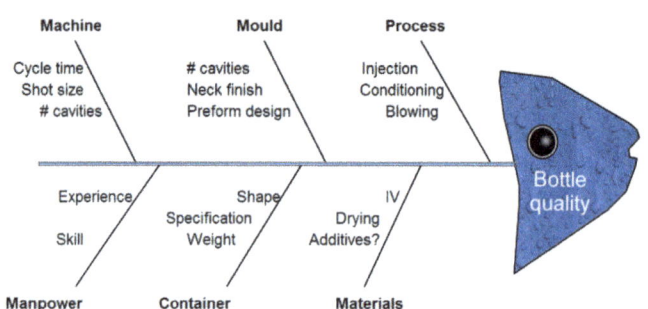

Good practice guidelines

- Verify for yourself what you have been told about the problem, any changes that have been implemented, and the outcome of checks already undertaken. Compare current production with previously approved samples.
- The optimum process settings should be recorded; ensure the machine is using them!
- Ask: "what has been changed (that might have triggered the fault)?" Or a variant on the same question: "when did the problem first start?" Changes made to the system are the most frequent cause of problems.
- Don't mistake coincidence for causality. Just because two events occurred at nearly the same time does not necessarily mean one event caused the other. Try duplicating the suspected cause to see if the fault reappears.
- Check to see whether other machines running the same material/ using the same dryer are experiencing similar problems.

---

[11] In order to clarify the issues, disparate but related ideas are grouped into meaningful categories that tie together different concepts having a common underlying theme (perhaps by sticking Post-it® Notes for each issue under appropriate headings on a large sheet of paper).

- If other machines exhibit similar problems, check whether those lines have anything in common, such as being supplied with chilled water or compressed air from the same source.
- If the condition is restricted to a single line, check to see whether it occurs in all cavities or just in particular cavities. Also determine whether the problem occurs every shot or appears randomly.
- Check first the most probable cause(s) of the problem/ things that are easiest to change. When making adjustments, try first those parameters that will give a quick result (whether beneficial or harmful).
- Make only one change at a time; ensure that the change is sufficient to produce a visible outcome, and allow time for the process to reach equilibrium before deciding upon the benefit (or otherwise) of the adjustment.
- Keep sample preforms and/ or containers and label them in order to maintain a record of process adjustments that have been tried.
- Exchange identical components and see whether or not the problem moves with the swapped component. If it does, you will have narrowed down the possible causes.
- Despite thermocouple and pressure gauge readings appearing normal, the measurements may not be reliable. If in doubt double check them with a separate instrument.
- Read the Manual! It really does make life easier if you are able to quickly turn to information about specific systems when it's needed.
- Focus on the problem rather than possible solutions, or to put it another way, don't jump to a solution before the problem is understood. If possible, take enough time to make sure that you understand what you are seeing and what is the root cause. Only by addressing the root cause of the problem will you be able to take the most effective corrective steps and ensure that the problem will go away permanently.
- Be aware that because the machine, the toolset, and the injection stretch-blow moulding process are inextricably linked, troubleshooting is likely to be an iterative procedure.
- After a long time spent troubleshooting a difficult problem, you are likely to become tired and may overlook crucial clues. Take your mind off the task at hand by having a break to allow your subconscious to work on the problem.
- Talk the problem over with someone else. This works best if you can discuss the problem with a person having intimate knowledge of the machine/ process, but simply outlining the problem to someone else can provide a fresh perspective.
- Once you have achieved a successful outcome, keep a permanent record of the symptoms and the solution in a log book. It could help you, or others, resolve future difficulties.
- When it is found necessary to change the process settings, maintain a written record of any alterations.

Electrical problems

Electricity can kill. Even straightforward tasks such as wiring a plug can lead to danger. Complicated tasks such as equipment repairs should only be tackled by people competent to do the job.

Electrical fault-finding starts from a proper understanding of how the equipment is designed to operate: it is much easier to analyse faulty operation when you know how it should operate. Machine manuals are vital sources of information and should be readily available.

- a high proportion of electrical and electronic system problems are caused by a very simple source of trouble: poor (i.e. open or shorted) connections. This is especially true when the environment includes such factors as high vibration. Bad connections in plug-and-socket connectors and terminal strips have the greatest risk for failure.
- tripped circuit breakers and blown fuses are other likely sources of trouble.
- remember that fuses burn out for a reason—find out why before replacing them.
- bear in mind that, due to their complexity, active components such as electro-mechanical relays tend to fail with greater regularity than passive devices such as proximity sensors.
- having identified the likely source of a problem, often the quickest method of pinpointing the faulty item is to swap it with an identical component.

Once the equipment has been repaired and put back in service, it is worth trying to determine the reason for the malfunction. Possible causes to consider include:

- did the component fail due to age?
- is there chafing that caused the wiring to short out (and how may this be prevented in future)?
- did the component fail due to improper use?
- might there be some other problem that is causing the same component to fail repeatedly?

Systematic troubleshooting procedure

If problems are not approached in a logical manner, time and effort can be wasted in locating the cause. A systematic procedure that narrows the problem down to an increasingly smaller area is more efficient than trial and error methods of locating faults. A logical troubleshooting procedure can be divided into three stages:

1 Name the fault
identify symptoms
don't waste time correcting acceptable 'faults' (the QA Standard should define limits)

2 Find the cause
list possible causes, use the symptom information along with knowledge of how the equipment is supposed to operate to make logical deductions

has something changed to trigger (or reveal) the fault ?

does the cause of the fault lie with the material, product (preform or container), process, machine, or mould ?

CHECK:
- *the mould (for damage)*
- *the material (correct grade, adequate dryness)*
- *the machine (correct temperatures and utility services, component malfunction ?)*
- *the conditions (process and machine settings)*

suggest possible remedies

3 Fix the fault
choose one remedy - consider short term variables first

SHORT TERM variables (results evident within a few cycles)
LONG TERM variables (results evident after a number of cycles)

EITHER:

alter process conditions
adjustments to conditions should be made methodically (one at a time) and efficiently (adjust those parameters that give the quickest results first) while the magnitude of the change should be sufficient to produce an easily visible result

observe and label the result. Identify and analyze trends

be aware that correcting one fault can produce another one

when process conditions have been altered, wait for equilibrium to be reached before altering other conditions to compensate for any undesirable changes that may have occurred

OR

repair / replace faulty component
operate machine to verify that the problem has been fixed

Preform faults & possible causes

Symptom	Possible Causes
Degraded material	Material too hot Screw recovery speed (RPM) too fast Barrel set temperature too high Hot Runner nozzles and manifold too hot Residence time too long
Preform sidewall haze	Low PET IV due to improper drying/ too much regrind Preform too hot Insufficient mould cooling
Crystallinity in gate area	Improper nozzle and/or manifold temperature Injection holding pressure too high
Holes in gate	Gate temperature too high Material cushion too small (screw may be *bottoming out*) Insufficient injection hold time / pressure Confirm Valve Gate shut-off pins are functioning correctly
Crystalline streaks	Hot runner manifold temperature too cold Injection speed too fast
Gate stringing	Gate temperature too high (or sometimes not high enough) Melt temperature too hot Injection holding pressure too high Confirm Valve Gate shut-off pins are functioning correctly
Bubbles	Material is melting too late, increase rear barrel temperature Screw recovery speed (RPM) too fast Back pressure too low Reduce decompression (screw retraction) time Change the size of the cushion Try a reverse (i.e. falling) barrel temperature profile Material not dry
Splay marks	*Defects that appear as smeared material or bubbles on the preform surface, usually oriented in the direction of flow* Remedy as for Bubbles
Cold slugs	Nozzle temperature too cold Confirm Valve Gate shut-off pins are functioning correctly
Unmelts	Melt temperature too cold Screw recovery speed (RPM) too slow Back pressure too low
Sink marks/ Short shots	Shot weight too low Injection pressure/ time too low Injection holding time too short Mould vents plugged, clean and check for damage Material cushion too small (screw may be *bottoming out*) Material too cold
Flashing	Injection pressure too high (changeover too late) Material too hot Mould in poor condition Insufficient clamp pressure

Preform faults & possible causes continued

Symptom	Possible Causes
Yellow preforms	Material dried too hot and/or too long Excessive amount of regrind
Weld lines in finish	Cavity vents plugged Injection speed too slow Material too cold (not knitting in weld)
Excessive shrinkage	Material temperature too high Inadequate holding pressure/ time Mould temperature too high
Poor concentricity	Injection Core pin bent or misaligned Injection speed too fast Taper surfaces worn
Preforms sticking/ distorting	Increase cooling time Confirm proper flow of chilled water Reduce chilled water temperature

Bottle faults and possible causes

Symptom	Possible Causes
Pearlescence	Preform too cold (in affected section)
Matt surface finish	Preform too hot Blow mould cooling inadequate
Thin sidewalls	Preform too hot IV too low Primary Blow air starting too late
Side to side problems	Preform too hot PET IV too low Stretch speed too slow Preform eccentricity (suggests a tooling problem)
Heavy necks	Preform too hot PET IV too low Primary Blow air starting too early
Heavy bases	Preform too hot PET IV too low Improper heating profile (too much heat in top area) Primary Blow air starting too late
Drop Test failure	Preform too hot Excessive gate crystallinity in preform
Top Load failure	Incorrect material distribution (see above symptoms)
Localized surface imperfections	Water in blow air Contaminated material Unmelts in preforms

12

Glossary

Glossary
an alphabetical list of words relating to a specific subject, text, or dialect, with explanations

ACETALDEHYDE (AA) is generated in small amounts during the melt processing of PET. Its presence in bottles results in an off-flavour in mineral waters and cola drinks. It is minimized by applying the mildest possible moulding conditions.

ACTUATOR - Component that converts hydraulic or pneumatic energy into mechanical energy.

AIR COMPRESSOR - A mechanical device that pressurizes gas in order to create power.

AIR RECEIVER - An air storage tank used with pneumatic systems that balances the air compressor capacity and airflow demand. Also known as an air reservoir.

BARRIER - A structure that prevents access into, or out of the container and thus protects the contents from deterioration in order to provide a longer shelf life. Generally this means preventing the penetration of oxygen and/ or moisture into the container, or preventing the carbon dioxide (CO_2) from exiting the container in the case of a carbonated soft drink.

BARRIER COAT - A surface coating to improve permeation resistance.

BASE - The bottom of the container, often marked with a material identification code to aid recycling, and sometimes bearing a registration indent used to orientate the container during labelling.

BIG BAGS - Form-stable flexible intermediate bulk containers (FIBC) are large fabric bags with lifting loops and fitted with a PE film liner that are used for the transport and storage of bulk materials such as plastic pellets.

BLOW MOLDING - A process in which a warm plastic parison or preform is placed between the two halves of a blow mould (cavity) and blown to form a container of the desired shape.

BLOW PRESSURE - The pressure required to form the parison or preform into the shape of the mould cavity in a blow moulding operation.

BODY - The principal part of a container consisting of the shoulders, sides (label panel) and base.

BOTTOM MOULD - That part of the blow mould which forms the heel radius and the base "push-up" of the container to be formed.

BRUCETON STAIRCASE - A statistical method of calculating the mean-failure height in a drop test.

BURNING is the appearance of brown or black smudges in the last region to fill. It occurs when the air cannot vent from the cavity rapidly enough to prevent it being compressed and thus heated.

CAD - Computer Aided Design

CAM - Computer Aided Manufacturing

CAPACITY - The amount of space for the product inside a container.

CAVITY - The part of the mould that contains the reverse image of the product being formed.

CSD - Carbonated Soft Drink

CHAMPAGNE BASE - Carbonated soft drink (pressure) bottle base design in the form of an inverted dome. How well it performs depends upon the thickness of the un-stretched amorphous material in the bottle base; the thicker it is, the greater the pressure it can support. The downside is that when amorphous PET is warmed, it softens, and will creep and deform under pressure. Also Champagne Bases performance decreases as the diameter of the inverted dome increases, which is why they tend to be used on smaller size bottles that taper in towards the bottom and which also have large heel radii. Has largely been supplanted by the Petaloid Base design.

CLOSURE - A term used to describe a metal or plastic cap which effects a primary seal when properly applied to a container.

CONCENTRICITY - The characteristic of circles, or circular cylindrical surfaces, of different radii having a common centre.

CONTINUOUS THREAD (C/T) FINISH - A continuous-thread finish that features an uninterrupted protruding helix on the neck of a container to accommodate a screw-type closure.

CONTROL VALVE - Mechanisms that control fluids in a pneumatic or hydraulic system. Control valves direct fluid movement and regulate the amount of pressure exerted in the system.

COPOLYMER - A material whose molecular chains are made from two different Monomers.

CORE - The part of an injection mould that forms the internal shaping of a product such as the internal threads of a closure.

CR - Child Resistant (of a closure)

CYCLE - The complete, repeating sequence of operations in a process. In moulding, the cycle time is the period of time between a certain point in one cycle and the same point in the succeeding one.

DECO LUG - A small indentation or raised portion on the surface of a container, usually in the base, that provides a means of positioning the container for processes such as pad printing.

DENSITY - Weight per unit of volume of a substance, expressed in grams per cubic centimetre, pounds per cubic foot, etc.

DIAPHRAGM ACTUATOR - A type of actuator that has a chamber divided in half by a diaphragm separating regions with different pressure levels.

DIMENSIONAL STABILITY - The ability of a material to maintain its shape under given processing or use conditions.

DROP TEST - A test of strength accomplished by dropping a container in a specified manner for a specified number of times, or until it fails from impact.

"E" DIMENSION - The external diameter of a threaded neck finish measured across the root of the threads (i.e. excluding the thread form).

ELASTIC STRAIN - Deformation that when released does not cause permanent deformation.

ENVIRONMENTAL STRESS CRACKING - The susceptibility of a plastic part to crack or craze when repeatedly stressed in the presence of certain chemicals.

FILL POINT - The level to which a container is designed to be filled.

FINISH (PLASTIC) - The neck of a bottle shaped to accommodate a specific closure. It carries the threads or lugs to which the closure is applied, and includes the sealing surface.

FLAME TREATMENT - A method of treating the container surface so as to render it receptive to printing inks and lacquers.

FLASH is excess plastic that has flowed out of the cavity and in between the mould faces where it has frozen to form thin, sheet-like protrusions around the parting line. It needs to be eliminated before the part can be considered acceptable.

FLEXIBILITY - The property of a material that permits it to be bent or twisted without breaking.

FLOWED-IN-GASKET - A gasket formed by a liquid material (vinyl or latex) poured (or flowed) directly into a gasket groove and cured in place, usually by baking; e.g. plastisol.

FLR-UNIT - A device that conditions air for use in pneumatic systems. A Filter-Lubricator-pressure-Regulator is commonly called an FLR-unit.

FLUORINATION - A surface treatment for polyethylene that is used to improve the barrier properties. (Not suitable for, and not required for PET).

GAS PERMEABILITY - The ability of a gas or other volatile substance to penetrate a material. Materials that will allow significant passage of gases are said to be permeable, while materials that resist or stop the passage of gases have Barrier properties.

GATE - The orifice through which the molten polymer enters the injection cavity.

GAYLORD - A generic term for a bulk box, a pallet-sized corrugated box.

HARDNESS - The resistance of a material to compression and indentation.

HAZE - A cloudy or foggy appearance in a normally transparent plastic.

"H" DIMENSION - The total height of the neck finish measured from the sealing surface to the neck bead or shoulder.

HEAD SPACE - The space between the level of the contents in the neck of a bottle and the closure. It enables the container to be opened without spillage, and also provides space for expansion of product due to heat or other action after filling.

HEAT EXCHANGER - Hydraulic component that relieves the excess heat that builds up in a hydraulic system.

HEAT SETTING a container involves raising the maximum working temperature of the oriented regions by blowing the preform into a heated blow mould. Drawbacks include that the blow moulds will be more expensive, there will be a cost for adding air blast cooling via a hollow stretch rod (essential when using very high blow mould temperatures), the blowing cycle time will be extended, and if used, special purpose heat-setting resins are generally more expensive to buy.

HEAT TRANSFER LABEL - A graphic that is reverse printed in the flat, usually in complex multi-colour, onto a film carrier. Heat transfers are used where multi-colour graphics are required.

HEEL - The part of a bottle between the bottle standing surface and the side wall (label panel).

HERMETIC SEAL - A seal that will exclude air and will be gas tight at normal temperatures and atmospheric pressures.

HOMOPOLYMER - A material whose molecular chains consist of one Monomer.

HYDRAULIC MOTOR - A device that converts the energy from a flowing liquid into mechanical motion.

HYDRAULIC PUMP - A mechanical device used to move liquids in a hydraulic system.

"I.D." - Abbreviation for inside diameter.

"I" DIMENSION - The internal diameter of the neck finish. Some closures require a precise "I" dimension.

INDENTED LABEL PANEL - When the diameter of the label panel area is smaller than the body diameter immediately above and below the label panel area.

INJECTION MOULDING - A moulding process in which, under pressure, plasticized polymer is forced from a cylinder into a cooled mould cavity to form a desired shape.

LINERLESS CLOSURE - One that has been engineered to function in specific applications without the use of an additional liner.

LABEL PANEL - The flat (often cylindrical) area of the container body intended to accommodate a label.

"L" DIMENSION - The vertical distance from the sealing surface to the top of the neck bead.

LIQUID COLOUR - A free-flowing dispersion of pigment in a polymer-compatible carrier which is introduced into the feed throat of the injection machine at a precisely controlled dose (typically by means of a peristaltic pump), where it mixes with the polymer pellets as they are plasticized to produce coloured mouldings. Liquid Colours can be started, stopped, or changed quickly and easily. (Oxygen scavenging and UV Barrier additives may be similarly added).

LUBRICITY - The property that diminishes friction and increases smoothness and slipperiness.

MANIFOLD - A pipe or chamber having multiple apertures for making connections.

MASTERBATCH - A high concentration of dye or pigment incorporated into plastic pellets. A strictly controlled amount is blended with the natural plastic material during plasticization in order to produce a precise, predetermined colour in the finished item. Both volumetric and gravimetric dosing systems are widely available.

MINIMUM WALL - A term that designates the minimum thickness of the wall of a container.

MONOMER - A simple organic molecule that can join in long chains with other molecules to form a macromolecule or Polymer.

MOUTH - The opening at the top of a container.

MULTI-LAYER BOTTLE - A bottle that is co-injected with two or more layers. The second material may have improved barrier properties. The process is used to make containers intended for oxygen sensitive foods or industrial chemicals. Multi-layer bottles can also be made with a layer of post-consumer resin (PCR) to incorporate recycled content.

NARROW NECK - A finish of a plastic container in which the diameter is small relative to the diameter of the body. Narrow mouth bottles are typically used for liquids and have neck sizes up to 28mm diameter.

NECK - The top of the container where the cross-section decreases to form the (neck) finish.

NECK FINISH - The neck of a bottle shaped to accommodate a specific closure. It carries the threads or lugs to which the closure is applied, and includes the sealing surface.

NECK INSERT - Part of the mould assembly that forms the neck and finish, also known as the "neck ring" or "neck splits".

NECK RING (SPLITS) - The part of the mould which forms the finish of a container.

NEWTONS PER SQUARE METER - A unit of pressure. A Newton per square meter is also known as a Pascal, which is derived from the International System of Units (SI).

OPAQUE - Descriptive of a material or substance which will not transmit light.

OVERFLOW CAPACITY - The maximum capacity of the container if it were to be filled to the very top of the finish.

PAD PRINTING - Used for printing across relatively small areas on plastic items, usually when the area to be printed is difficult to access or the surface is contoured.

PANELLING - Distortion or sidewall collapse of a container occurring after filling during storage. Panelling is typically the result of reduced pressure inside the container.

PARISON - The extruded tube of hot plastic that will be placed in a blow mould to be inflated into a bottle or other hollow form.

PARTING LINE - The mark on a container where the two halves of the blow mould meet in closing.

PERMEATION - The extent to which a gas or water vapour passes through a plastic film or container. Permeability: the rate of such passage.

PETALOID BASE - Carbonated soft drink (pressure) bottle base design with five lobed feet.

PINCH-OFF - A raised edge around the cavity in the blow mould which seals off the part and separates the excess material as the mould closes onto the parison in the extrusion blow moulding process.

PLASTIC MEMORY - The tendency of plastics to return to their original moulded form when heated.

PLASTICIZATION - Conversion of the solid polymer into a liquid melt inside the injection cylinder. The polymer pellets are fed from a Hopper into the cylinder where they are melted as they are forced forward along the inside of the barrel by the rotation of the screw.

PNEUMATIC SYSTEM - A power transmission system that uses the force of flowing gases to transmit power.

POLYMER - A compound that consists of long macromolecules made up of many chemically bonded small identical molecules strung together.

POLYOLEFIN - A polymer produced from a simple olefin monomer with the general formula $C_n H_{2n}$. Well known polyolefins used for making containers are HDPE and PP.

POST-CONSUMER RESIN (PCR) - Recycled plastic which after a previous life as a container, has been sorted, cleaned, and reprocessed so that it may be re-used to make new containers.

PREFORM - In Blow Moulding a thermoplastic polymer such as PET is formed into a shape similar to a thick-walled test tube with a thread form around the opening known as a Preform. The hot Preform is then inflated inside a Blow Mould to create a container of the desired shape.

PRESSURE-RELIEF VALVE - A control valve that opens when the set fluid pressure is exceeded.

PRESSURE-SENSITIVE LABELS - Self-adhesive labels which are carried on roll stock, and applied to a container or closure in a manual or automatic process.

PROTOTYPE MOULD - A simplified mould construction sometimes made from a light-casting alloy or even epoxy resin in order to obtain information for the final mould or part design.

PUSH UP - The recess in the bottom of a plastic container designed to provide a flat standing surface around its circumference so that the container does not exhibit a rocker bottom.

REGISTRATION RECESS (SEAT) - A small indentation in the surface of a container, usually in the base, which provides a means of positioning the container for automated processes such as multi-colour decoration or labelling.

REGRIND - Thermoplastic material from a processor's in-house scrap production that has been ground into flakes so that it may be blended with virgin material and used to make new product.

ROTARY ACTUATOR - A valve actuator that converts fluid flow into circular motion.

RUNNER - In injection moulding, one of the passages that convey plastic melt from the injection point (sprue) and distribute it to the various cavities in a multi-cavity mould.

SCREEN PRINTING (also called Silk Screen Printing) - A printing technique that involves squeezing the printing ink through a fabric stretched on a frame onto which a stencil has been applied. The stencil openings determine the shape and density of the ink transferred to the container during the screen printing process.

SCREW COMPRESSOR - An air compressor which has meshed gears that rotate to move air through a pneumatic system.

"S" DIMENSION - The distance from the top of the neck finish to the top edge of the first full thread.

SEAT (REGISTRATION RECESS) - Indentation on the base of a container that is used to align it on filling and decorating lines so that it is correctly oriented.

SHELF LIFE - The period of time during which a product can be stored under specified temperature and humidity conditions and remain suitable for use. Shelf life is also sometimes called storage life.

SHORT SHOTS are shots that do not fill the mould completely. The plastic has simply not flowed far enough to fill the part.

SHOULDER - The conical region of a bottle between the neck and the body.

SHRINKAGE (MOULD) - The decrease in size a moulded article undergoes after being moulded. Shrinkage is cause by cooling and subsequent contraction of the plastic material.

SPLAY - A surface defect usually caused by trapped gas bubbles being smeared across the surface as the flow front moves to fill the part. The bubbles can be due to the presence of moisture in the material (i.e. inadequate drying), or air entrapment.

SPRAY COATING - A PVDC film applied to PET that improves Barrier properties.

STACK HEIGHT - Distance from the top of one container to the top of another standing on top of it.

STRETCH ROD - Used in Injection Stretch Blow Moulding. A rod that is introduced into the Preform to stretch it in an axial direction at the same time as the preform is being blown.

TAMPER-EVIDENCE - Any device that provides visible evidence that a container has been opened.

"T" DIMENSION - The outside diameter of the thread helix on a neck finish.

T_g - Glass to Rubber transition temperature.

THERMAL STABILITY - The ability of a container to maintain its shape under given conditions.

THERMOFORMING - A method of forming plastics in which plastic sheet material is heated to a point where it is soft and pliable. The sheet is then formed to the desired shape using vacuum, pressure and mechanical assists, or any combination of these.

THERMOPLASTIC - Material that will repeatedly soften when heated and harden when cooled.

THREADS - On continuous thread styles, the spiral of plastic onto which a C/T closure is twisted. Different C/T closure styles feature different numbers of threads.

TORQUE - Twisting force used to attach or remove the closure.

U.V. (ULTRAVIOLET) INHIBITOR - A chemical added to a plastic resin, which absorbs UV light and helps prevent damage to, and prolongs the life of, the plastic.

U.V. STABILIZER - Any chemical compound which, when admixed with a thermoplastic resin, selectively absorbs UV rays and minimizes chemical and/ or physical changes that may be caused.

UNIT CAVITY - A mould with only one cavity, usually a pilot for the production set of moulds.

VALVE MANIFOLD - A standard mounting block used in fluid power systems that simplifies the use of multiple control valves.

VOLUME - Also referred to as displacement or capacity. (1) The fill volume is the amount of water a bottle is designed to hold up to the fill point of the bottle. (2) The over flow capacity is the amount of water a bottle will hold when filled to overflowing.

WALL THICKNESS - The thickness of the bottle wall.

WELD LINES - A visible line created in the surface of a moulded part caused by two plastic flow fronts joining inside the cavity as it fills.

WIDE MOUTH - Container with large diameter neck opening greater than 28mm, or containers that have a large finish size in relation to their capacity.

YIELD POINT - That point beyond which the stresses applied to a material will cause permanent deformation.

Conversion table

multiply by factor →

	Unit	Abbreviation	Factor	Abbreviation	Unit
Force	Kilogram force	kgf	9.80665	N	Newton
Force	Pound force	lbf	4.44822	N	Newton
Work. energy	Foot pound	ft lbf	1.35582	J	Joule
Power	Horsepower	hp	0.7457	kW	Kilowatt
Power	Metric horsepower	PS	0.735499	kW	Kilowatt
Power	Foot pound/minute	ft lbf/min	81.3492	W	Watt
Torque	Kilogram force metre	kgf m	9.80665	Nm	Newton metre
Torque	Kilopond metre	kp m	9.80665	Nm	Newton metre
Torque	Pound force. foot	lbf ft	1.35582	Nm	Newton metre
Torque	Pound force. inch	lbf in	0.112985	Nm	Newton metre
Pressure	Atmosphere	Atm	1.013250	bar	Bar
Pressure	Kilogram force/sq centimetre	kgf/cm^2	0.980665	bar	Bar
Pressure	Kilopascal	kPa	0.01	bar	Bar
Pressure	Newton/square centimetre	N/cm^2	0.1	bar	Bar
Pressure	Newton/square metre	N/m^2	0.00001	bar	Bar
Pressure	Pascal (Newton/sq meter)	Pa	0.00001	bar	Bar
Pressure	Pound force/ square inch	lbf/in^2	0.06894	bar	Bar
Pressure Water column	Inch of water	in H$_2$O	2.49089	mbar	Millibar
Pressure Water column	Foot of water	ft H$_2$O	0.0298907	bar	Bar
Pressure Water column	Millimetre of water	mm H$_2$O	0.09806	mbar	Millibar
Pressure Mercury column	Inch of mercury	in Hg	33.8639	mbar	Millibar
Pressure Mercury column	Millimetre of mercury	mm Hg	1.33322	mbar	Millibar
Degree of angle	Degree (angle)	°	0.0174533	rad	Radian
Length	Foot	ft	0.3048	m	Metre
Length	Inch	in	25.4	mm	Millimetre
Area	Square foot	ft^2	0.092903	m^2	Square metre
Area	Square inch	in^2	0.000645	m^2	Square metre
Area	Square inch	in^2	6.4516	cm^2	Square centimetre
Volume	Cubic centimetre	cm^3	0.001	l	Litre
Volume	Cubic centimetre	cm^3	1.0	ml	Millilitre
Volume	Cubic foot	ft^3	0.028317	m^3	Cubic metre
Volume	Cubic foot	ft^3	28.3168	l	Litre
Volume	Cubic inch	in^3	16.3871	cm^3	Cubic centimetre
Volume	Cubic inch	in^3	0.016387	l	Litre
Volume	Fluid ounce. UK	UK fl oz	28.4131	cm^3	Cubic centimetre
Volume	Fluid ounce. US	US fl oz	29.5735	cm^3	Cubic centimetre
Volume	Gallon. UK	UK gal	4.54609	l	Litre
Volume	Gallon. US	US gal	3.78531	l	Litre
Volume	Pint, UK	UK Pt	0.568261	l	Litre
Volume	Pint, US	US liquid pt	0.473176	l	Litre
Mass	Pound (mass)	lb	0.45359	kg	Kilogram
Density	Pound/cubic inch	lb/in^3	0.0276799	kg/cm^3	Kilogram/cubic centimetre
Density	Pound/cubic foot	lb/ft^3	16.0185	kg/m^3	Kilogram/cubic metre
Temperature	Fahrenheit	°F	*	°C	Degree Celsius
Quantity of heat	BTU/hour	Btu/h	0.293071	W hr	Watt hours
Rotational frequency	Revolutions/minute	rpm	0.104720	rad/s	Radians/second

* °C = 5 x (°F - 32) / 9

← *divide by factor*

Apparent Density of Water in Air

Temperature °C	Density [grams/ml]	Temperature °F
0	0.99873	32.0
1	0.99879	33.8
2	0.99884	35.6
3	0.99887	37.4
4	0.99888	39.2
5	0.99887	41.0
6	0.99885	42.8
7	0.99882	44.6
8	0.99877	46.4
9	0.99871	48.2
10	0.99863	50.0
11	0.99854	51.8
12	0.99843	53.6
13	0.99832	55.4
14	0.99819	57.2
15	0.99805	59.0
16	0.99790	60.8
17	0.99773	62.6
18	0.99756	64.4
19	0.99737	66.2
20	0.99717	68.0
21	0.99697	69.8
22	0.99675	71.6
23	0.99652	73.4
24	0.99628	75.2
25	0.99603	77.0
26	0.99578	78.8
27	0.99551	80.6
28	0.99523	82.4
29	0.99495	84.2
30	0.99465	86.0
31	0.99435	87.8
32	0.99404	89.6
33	0.99372	91.4
34	0.99339	93.2
35	0.99305	95.0

The above density values as used by the Coca-Cola Company, are calculated from the 58th Edition of the Handbook of Physics and Chemistry. page F-2 (Weight of 1 Gallon of Water) in which the weights are for dry air at the same temperature as the water. While the number of decimal places might suggest that the density of water is an absolute, as a matter of fact it is not only the temperature and ambient humidity which influences the density, but also whether the water is drawn straight from the tap, or has first been distilled or deionised.

Neck finish terminology

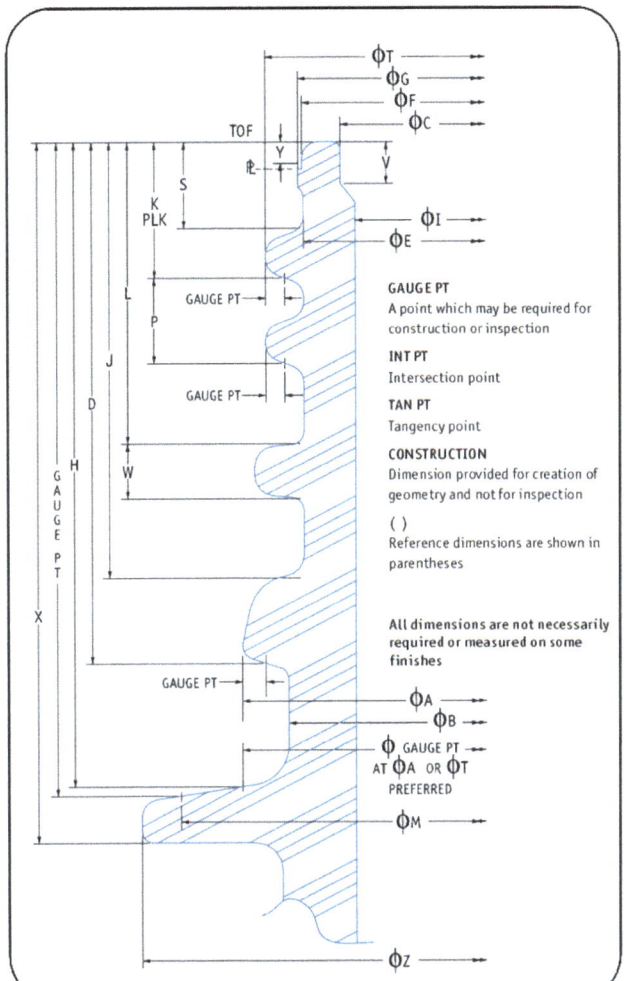

GAUGE PT
A point which may be required for construction or inspection

INT PT
Intersection point

TAN PT
Tangency point

CONSTRUCTION
Dimension provided for creation of geometry and not for inspection

( )
Reference dimensions are shown in parentheses

All dimensions are not necessarily required or measured on some finishes

TFRSC Nomenclature

- A – Tamper evident bead diameter
- B – Tamper evident band recess diameter
- ß – Helix angle at pitch diameter
- C – Control diameter at TOF
- D – Tamper evident bead height measured from TOF to gauge pt
- E – Thread root diameter
- F – Upper ring diameter
- G – Lower ring diameter
- H – Clearance height required for proper closure function
- I – Minimum diameter through finish
- J – Height from TOF to the top of the tamper evident bead
- K – Height from TOF to the gauge pt at the start of full thread (at thread start position)
- L – Height From TOF to top of bead
- M – Construction gauge pt on support ledge
- P – Thread pitch (measured from K)
- ₧ – Parting-line
- PLK– Height from TOF to the gauge pt of first full thread taken at parting-line (clockwise from thread start position)
- S – Height from TOF to start of full depth of thread
- T – Thread crest diameter
- TOF – Top of finish
- V – Control diameter depth
- W – Width of bead
- X – Height from TOF to bottom of support ledge
- Y – Seal control length
- Z – Maximum diameter on support ledge

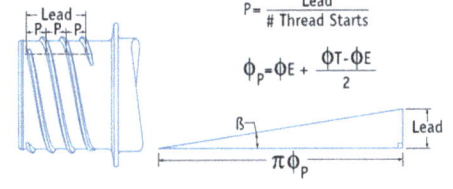

$$P = \frac{Lead}{\text{\# Thread Starts}}$$

$$\phi_P = \phi E + \frac{\phi T - \phi E}{2}$$

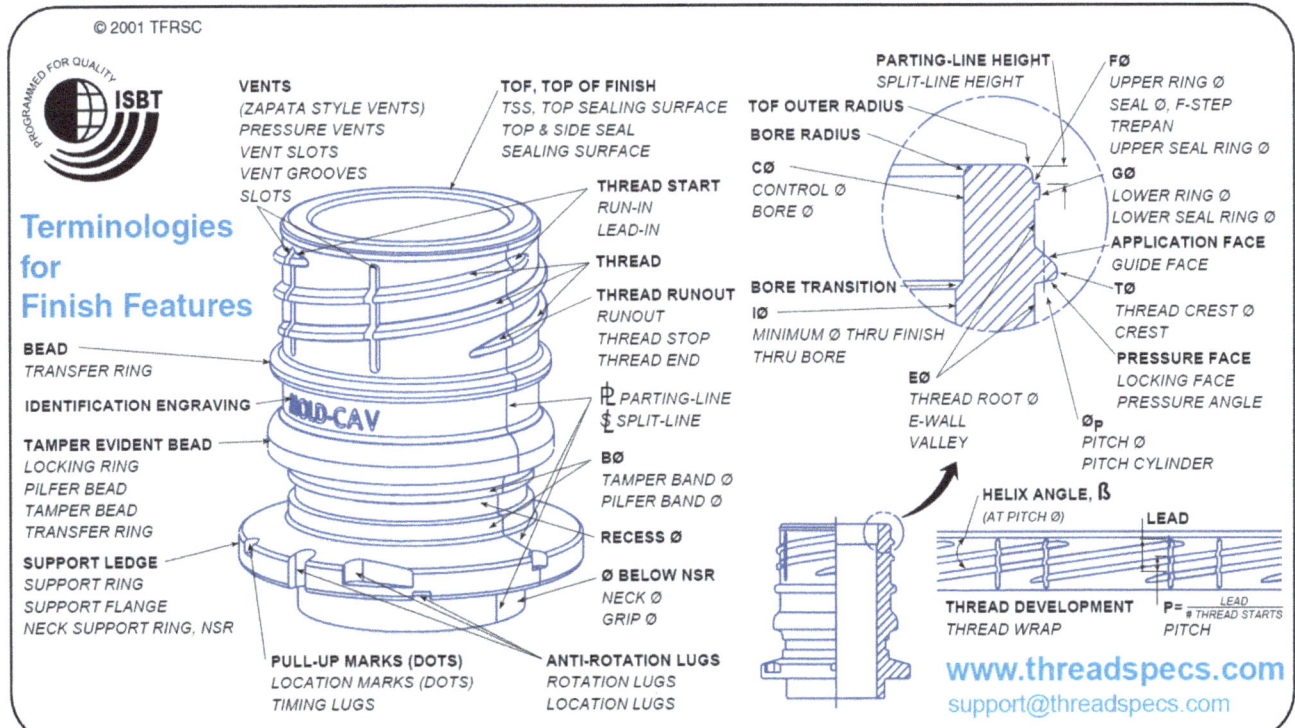

Colour

Amorphous PET is not completely transparent; catalyst residues from the production process and reheat additives if present, tint the resin. Material data sheets specify the colour in terms of the CIE L* a* b* colour space.

The International Commission on Illumination (usually abbreviated CIE for its French name, *Commission internationale de l'éclairage*) is an international body of colour scientists based in Vienna whose standards make it possible to precisely communicate colour information. The CIE L* a* b* colour space uses three values or coordinates to describe a colour. **L** describes relative lightness and extends from 0 (black) to 100 (white), **a** represents relative redness–greenness, and **b** represents relative yellowness–blueness.

The colour axes are based on the fact that a colour can't be both red *and* green, or both blue *and* yellow at the same time, because these colour pairs oppose each other. On each axis the values run from positive to negative. On the a−– a+ axis, positive values indicate amounts of red while negative values indicate amounts of green. On the b−– b+ axis, yellow is positive and blue is negative. For both axes, the zero point is neutral grey. To define any colour, values are only needed for each of the two colour axes, and the lightness (or darkness), which is on the separate L* axis, where 0 = black, and 100 = white

+a is the redness, and -a is the greenness

+b is the yellowness, and -b is the blueness

Samples for which a* = 0 and b* = 0 are achromatic (without colour), and thus the L* axis represents the achromatic scale of greys from black to white.

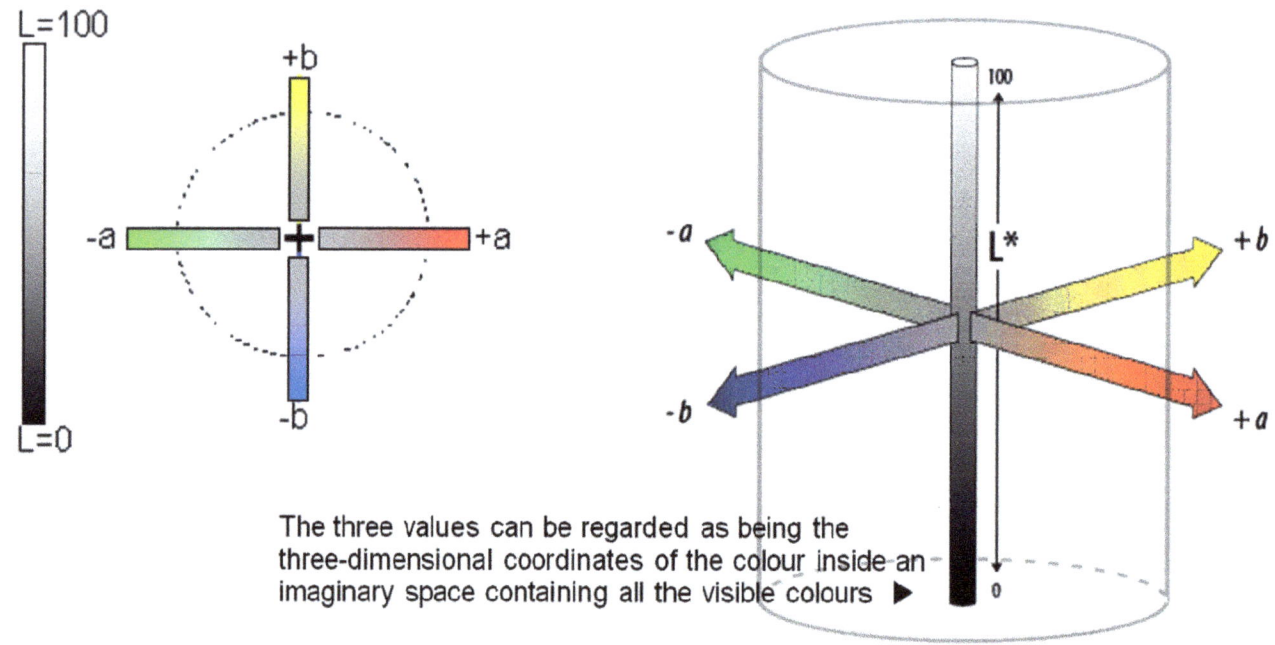

The three values can be regarded as being the three-dimensional coordinates of the colour inside an imaginary space containing all the visible colours ▶

INDEX

AC solenoid, 75
Acetaldehyde, 36, 94
Actuator (electric), hydraulic valve, 53
Air shot, 18
Amorphous PET, 30, 31, 34
Annual output per machine, 80
Anti-rust Spray, 87
Anti-seize compound, 23
Automatic wall thickness control, 24
Axial stretch ratio, 36, 37
Back pressure, 17
Back pressure (injection), 105
Balanced fill, 20
Balayage, 38
Barrier Screw, 16
Biaxial orientation, 11, 32
Blow air, example calculation, 84
Bottle faults and possible causes, 120
Bottle tests, 91
Burst pressure of container, 94
Capacity tolerance for containers, 92
Catalyst residues, 34
Centralised material drying system, 79
Chain-coupling (hydraulic pump), 49
Champagne base, 109
Check Ring, 14
Check valve (hydraulic), 56
Chilled water circulation (cores & cavities), 105
Clamping Unit, 19
Commodity product (PET), 34
Compressed air, 81
Compressed Air Reservoir, 83
Conditioning station, 14, 23
Container wall thickness measurement, 92
Contamination in hydraulic systems, 61
Continuous process control, 98
Control Charts, 96
Cooling systems, Factory, 81
Cooling time (injection), 105
Co-polymer, 34
Crater (blowing issue), 108
Creep (long-term thermal expansion), 37, 95
Crystalline PET, 30, 32
Cushioning (pneumatic actuators), 66
Daylight, 19
DC solenoid, 75
Dehumidifying dryer, 45
Directional valve (pneumatic), 66
Directional valves (hydraulic), 53
Double-acting pneumatic cylinder, 64
Drooling, minimize (injection), 105
Drop testing of containers, 93
Drying, temperature and duration, 42, 43, 45
Duty cycle, Air Compressor, 82
Electrical systems and schematics, 69, 77
Emergency stop button, 75
Energy efficiency, 77

Extrusion blow moulding, 9
Factory layout, 79
Filter (hydraulic oil), 48
Flake (regrind PET), 36
Flash, 18, 20
Floor loading, 80
Flow balanced hot runner system, 21
Flow Control Valve (hydraulic), 56
Flow restrictor (hydraulic), 54
Gate freeze/ close time determination, 104
Gate orifice, 21
Glass transition temperature T_g, 31, 43
Grease, food grade, 62
Health and Safety at Work Act, 7
Heat exchanger (hydraulic oil), 48
Heat Setting, 37
Heater bands, 18, 73
Heating Pot, 23
Hoop stretch ratio, 36, 37
Hot fill capability, 38, 95
Hot runner, 20
Hot runner nozzle, 21
Hydraulic Cylinder, 61
Hydraulic pump, 49
Hydraulic system, 47
Hydrolysis, 35, 41
Hygroscopic, 35, 41
Induction motor, 70
Injection Blow Moulding, 10
Injection capacity, 16
Injection cylinder temperature profile, 102
Injection mould, 19
Injection pressure, 16, 103
Injection screw, 15
Injection speed, 17, 104
Injection stretch-blow moulding, 9
Injection stroke, 16
Injection time, 104
Injection unit, 14
Intensification ratio, 16, 19
Intrinsic viscosity (IV), 33, 41, 91
L/D ratio, 15
Ladder diagram (PLC program), 72
Leak testing of containers, 93
Lighting, Factory, 80
Limit switches, 75
Maintenance (hydraulic), 61
Manual override (hydraulic valve), 54
Materials, polymeric, 29
Melt decompression, 22
Melt strength, 34
Meter-in hydraulic circuit, 61
Meter out hydraulic circuit, 61
Moisture Vapour Transmission Rate, 94
Mollier Diagram, 44
Mould opening stroke, 19
Natural stretch ratio (PET), 32, 33

Neck crystallisation (PET bottle), *39*
No Fuse circuit breaker, *75*
Nomex, *23*
Oil motor, *58*
Oil tank (hydraulic reservoir), *47*
Oil-free compressed air, *25*
One-stage process, *10*
Operation air, *64*
Optimisation of ISBM process, *101*
Orientation, biaxial, 11, *31*
Overall Equipment Effectiveness (OEE), *98*
Oxygen permeation through container wall, *94*
Packing pressure (injection), *14*, *104*
Parts-per-million fractions, *42*
Pearlescence (preform temperature issue), *108*
PET regrind, *35*
Photo-sensor, *76*
PID temperature controller, *74*
Pilot Operated 2-Way Valves, *26*
Pilot-operated hydraulic relief valve, *51*
Plasticizing capacity, *16*
Pneumatic cylinder, *64*
Pneumatic systems and symbols, *63*
Polarised light inspection, *91*
Polyethersulfone, *40*
Polymer, *29*
Polypropylene, *39*
Preform design, *36*
Preform faults & possible causes, *119*
Preform neck dimensions, *91*
Preform tests, *90*
Preform, uneven wall thickness, *91*
Pressure balanced, hot runner system, *21*
Pressure reducing valve (hydraulic), *54*
Pressure regulator, pneumatic, *64*
Pressure Relief valve (hydraulic), *51*
Primary blow air pressure, *108*
Problem solving techniques, *116*
Programmable Logic Controller, *71*
Proportional valves, *58*
Proximity sensor, *66*, *75*, *117*
Push button switches, *75*
Quick exhaust valve (pneumatic), *67*
Reciprocating air compressor, *82*
Reciprocating screw, *14*
Refrigerated air dryer, *83*
Regeneration circuit, *45*
Regenerative hydraulic circuit, *56*
Regrind PET, *35*
Reheat additives (two-stage), *35*
Relative Humidity, *42*
Relative Humidity (RH), *33*
Relays (electrical system), *73*
Residence time (cylinder), *14*, *16*
Residence time (dryer), *44*
Returnable/ refillable PET bottle tests, *96*
Rotary Actuator, *60*
Rotation reversal, induction motor, *70*

Runners, hot runner system, *20*
Sacrificial anode (hydraulic reservoir), *48*
Safety interlock switches, *76*
Screw air compressor, *82*
Screw (injection) rotation speed (charging), *102*
Screw rotation speed—RPM, *17*
Screw surface speed, *17*
Secondary blow air pressure and time, *108*
Selector switch, *75*
Self healing effect, *32*, *34*
Shear heating, *103*
Shooting pot, *17*
Short shot, *103*
Shot size (injection), *103*
Shrinkage of containers after blowing, *91*
Shrinkage, polymer, *31*
Shut height, *19*
Single-phase electricity, *69*
Sink marks, *18*
Solenoids (electrical system), *75*
Spare parts Store, *87*
Speed Control Valve, pneumatic, *67*
Splay marks, *106*
Spool valve, *53*, *66*
Sprue Bush, *14*, *17*, *20*
Staffa hydraulic motor, *58*
Star-Delta starter, *71*
Statistical process control, *96*
Stator (induction motor), *70*
Stretch ratio, *32*
Stretch Rod, *24*, *107*
Stretch-blow moulding, *24*
Stringing, *21*, *106*
Suction filter (hydraulic reservoir), *47*
Tacky (sticky PET), *43*
Tainting (acetaldehyde), *36*
Take-out station, *14*
Temperature Controllers, *74*
Thermal break hot runner nozzle, *21*
Thermal expansion, *22*
Thermal overload relay, induction motor, *71*
Thermal stability, *95*
Thermocouples, *73*
Thermoplastic and thermosetting polymers, *29*
Timers (PLC program), *72*
Timing Charts, 109
Transformer, electrical, *70*
Troubleshooting, *115*
Troubleshooting, systematic procedure, *118*
Turbulent flow, *20*
Two-stage process, *10*
Under-packed preforms, *103*
Unloading valve (hydraulic), *55*
Vacuum strength of containers, *94*
Valve gate nozzle, *21*, *22*
Vented injection cylinder (barrel), *46*
Voltage Regulator, *75*
White food grade grease, *62*

Notes

135

www.ingramcontent.com/pod-product-compliance
Lightning Source LLC
Chambersburg PA
CBHW041545220526
45473CB00014B/2957